Site Layout Level One

PEARSON

Prentice
Hall

Upper Saddle River, New Jersey
Columbus, Ohio

contren®
Learning Series

National Center for Construction Education and Research

President: Don Whyte
Director of Curriculum Revision and Development: Daniele Dixon
Site Layout Project Manager: Rebecca Hassell
Production Manager: Debie Ness
Quality Assurance Coordinator: Jessica Martin
Editor: Bethany Harvey
Desktop Publisher: Laura Parker, Debie Ness

The NCCER would like to acknowledge the contract service provider for this curriculum:
Topaz Publications, Liverpool, New York.

This information is general in nature and intended for training purposes only. Actual performance of activities described in this manual requires compliance with all applicable operating, service, maintenance, and safety procedures under the direction of qualified personnel. References in this manual to patented or proprietary devices do not constitute a recommendation of their use.

10 9 8 7
ISBN 978-0-13-109173-3

PEARSON
Prentice
Hall

nccer

contren·
Learning Series

Preface

This volume was developed by the National Center for Construction Education and Research (NCCER) in response to the training needs of the construction, maintenance, and pipeline industries. It is one of many in NCCER's *Contren® Learning Series*. The program, covering training for close to 40 construction and maintenance areas, and including skills assessments, safety training, and management education, was developed over a period of years by industry and education specialists.

NCCER also maintains a National Registry that provides transcripts, certificates, and wallet cards to individuals who have successfully completed modules of NCCER's *Contren® Learning Series*, when the training program is delivered by an NCCER Accredited training Sponsor.

The NCCER is a not-for-profit 501(c)(3) education foundation established in 1995 by the world's largest and most progressive construction companies and national construction associations. It was founded to address the severe workforce shortage facing the industry and to develop a standardized training process and curricula. Today, NCCER is supported by hundreds of leading construction and maintenance companies, manufacturers, and national associations, including the following partnering organizations:

PARTNERING ASSOCIATIONS

- American Fire Sprinkler Association
- American Petroleum Institute
- American Society for Training & Development
- American Welding Society
- Associated Builders & Contractors, Inc.
- Association for Career and Technical Education
- Associated General Contractors of America
- Carolinas AGC, Inc.
- Carolinas Electrical Contractors Association
- Citizens Democracy Corps
- Construction Industry Institute
- Construction Users Roundtable

- Design-Build Institute of America
- Merit Contractors Association of Canada
- Metal Building Manufacturers Association
- National Association of Minority Contractors
- National Association of State Supervisors for Trade and Industrial Education
- National Association of Women in Construction
- National Insulation Association
- National Ready Mixed Concrete Association
- National Systems Contractors Association
- National Utility Contractors Association
- National Vocational Technical Honor Society
- North American Crane Bureau
- North American Technician Excellence
- Painting & Decorating Contractors of America
- Plumbing-Heating-Cooling Contractors National Association
- Portland Cement Association
- SkillsUSA
- Steel Erectors Association of America
- Texas Gulf Coast Chapter ABC
- U.S. Army Corps of Engineers
- University of Florida
- Women Construction Owners & Executives, USA

Some features of NCCER's *Contren® Learning Series* are:

- An industry-proven record of success
- Curricula developed by the industry for the industry
- National standardization providing portability of learned job skills and educational credits
- Credentials for individuals through NCCER's National Registry
- Compliance with Apprenticeship, Training, Employer, and Labor Services (ATELS) requirements for related classroom training (CFR 29:29)
- Well-illustrated, up-to-date, and practical information

Acknowledgments

This curriculum was revised as a result of the farsightedness and leadership of the following sponsors:

Paul A. Castaneda
Chuck Hogg
Ken McGhee
Dan Musselwhite
Steven C. Schipper
Verne Shaw

This curriculum would not exist were it not for the dedication and unselfish energy of those volunteers who served on the Authoring Team. A sincere thanks is extended to:

Cianbro Corporation
Holder Engineering Services
Sundt Construction Inc.

Contents

Introduction to Site Layout

COURSE MAP

This course map shows all of the modules in the first level of the Site Layout curriculum. The suggested training order begins at the bottom and proceeds up. Skill levels increase as you advance on the course map. The local Training Program Sponsor may adjust the training order.

SITE LAYOUT LEVEL ONE

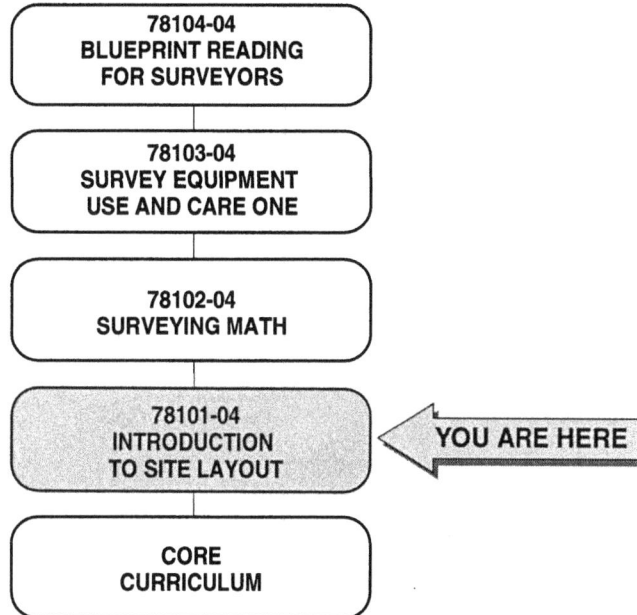

78104-04
BLUEPRINT READING
FOR SURVEYORS

78103-04
SURVEY EQUIPMENT
USE AND CARE ONE

78102-04
SURVEYING MATH

78101-04
INTRODUCTION
TO SITE LAYOUT ← YOU ARE HERE

CORE
CURRICULUM

101CMAP.EPS

MODULE 78101-04 CONTENTS

Figures

Introduction to Site Layout

Objectives

When you have completed this module, you will be able to do the following:

1. State the purpose of site layout and describe the role of a site layout technician in the construction industry.
2. Describe the different kinds of surveys related to a construction project.
3. Explain the relationship between Earth's latitude and longitude lines and how these lines are used.
4. Define the various survey control points used in the process of site layout.
5. Explain the meaning of terminology used in site layout.
6. Identify the career opportunities available to people in the site layout field.
7. State the rules for the professional and ethical conduct of a site layout person.
8. Set up a builder's level, shoot an elevation, and properly record the data.
9. Take an inverted rod reading.
10. Chain a distance on even terrain and correctly record the data.
11. Use a plumb bob correctly.
12. Set up a tripod correctly.

Prerequisites

Before you begin this module, it is recommended that you successfully complete the following: Core Curriculum.

1.0.0 ◆ INTRODUCTION

Site layout is an important phase of any construction activity from beginning to end (*Figure 1*).

The impact of good layout practices in terms of time, money, and construction efficiency increases with the size of the project. Site layout technicians typically survey and otherwise gather information about a proposed job site. As a job progresses, they lay out the location of structures and other features on the site, check the dimensions of the structures as they are being built, document the completed work, and verify that all the work is done in accordance with the design plans and specifications.

101F01.EPS

Figure 1 ◆ Typical construction site.

As a construction project progresses, numerous activities involve site layout tasks. These activities typically can include the following:

- Evaluate the site topography to determine if it is level or if fill or excavating is needed
- Establish test points for soil and subsurface testing
- Set up **control points** using appropriate monuments and markers
- Set up silt and tree-save fences

- Set stakes for areas to be cleared of stumps, brush, and other debris
- Set up access roads
- Locate and plan utilities
- Lay out soil removal/grading
- Set up a storm water plan
- Stake out power
- Locate job-site facilities for construction trailers, fuel storage, and other needs
- Lay out footings, foundations, and pilings
- Lay out structural steel and structural concrete and metal decks
- Lay out columns and beams
- Monitor rebar and concrete placement
- Perform post tension inspections
- Set final control lines on floor, deck, or slab
- Verify all slab elevations

2.0.0 ◆ SITE LAYOUT TERMINOLOGY

As with any construction trade, numerous trade terms that are unique to the site layout tasks and trade are used. To be successful in site layout, you must understand these terms. Throughout this curriculum, trade terms are introduced and used whenever appropriate to facilitate the instruction. When a new trade term is introduced in the curriculum for the first time, it is indicated by boldface print in the module text and further explained in the glossary presented at the back of the same module. As a result, upon completion of your studies in the *Site Layout* curriculum, you will have been introduced to and understand many of the commonly used trade terms pertaining to site layout.

3.0.0 ◆ SURVEYING OVERVIEW

Surveying is performed almost daily by site layout technicians. It involves determining the location of physical land and/or man-made features relative to one another and relative to a predetermined reference on the surface of Earth. Surveying is considered scientific in nature because it involves mathematics to analyze and adjust the survey data. Field survey measurements include horizontal and slope distances, vertical distances, and horizontal and vertical angles. Surveys can be divided into several categories and types, the most common of which are described here.

3.1.0 Categories of Surveys

The two main categories of surveys are **geodetic surveys** and **plane surveys**. Geodetic surveys involve the surveying of large areas over long distances, such as those involved with state boundaries and international borders. Because the size of the geographical area involved in a geodetic survey is so large, the curved shape of Earth's surface must be taken into account and the measured horizontal distances and angles adjusted accordingly. Height (elevation) dimensions for geodetic surveys are referenced to a known **datum**, usually **mean sea level (MSL)**. MSL is assigned the elevation of 0.000 feet (or meters), and all points on Earth can be described by elevations above and below zero. Numerous permanent elevation datum points (**bench marks**) located throughout the United States have precisely determined elevations available for survey use. Geodetic surveys are typically performed by government agencies. They are carried out with great precision and are used to establish highly accurate control networks. Control networks are covered later in this module.

Plane surveys involve relatively small geographical areas, usually no more than a few square miles. Most construction site layout projects fall in the category of plane surveys. Because of the small area of Earth's surface being surveyed, Earth can be treated as a horizontal flat, or plane, surface. This allows the effects of Earth's curvature to be ignored when making surveying distance and angular measurements and computations. As with geodetic surveys, elevation dimensions are referenced to a known datum. However, the datum used is usually a locally established or arbitrarily selected elevation point other than mean sea level.

3.2.0 Types of Surveys

Several different types of surveys are performed in support of a particular construction project as it progresses from start to finish. These types of surveys include the following:

- Control surveys
- Preliminary surveys
- Construction surveys
- As-built surveys

3.2.1 Control Surveys

Control surveys establish the network of horizontal and vertical reference points needed to perform subsequent preliminary and construction surveys. During the control survey, horizontal (**line**) and vertical (**grade**) positions of a number of points, called control points, are accurately established at the boundaries of and within the area being surveyed. For a typical project, the horizontal control points are usually tied to property lines, to the center line of a road, or to some other permanent feature. The vertical control points (bench marks) are a permanent set of points of known elevation above or below sea level. Note that control points are discussed in more detail later in this module.

3.2.2 Preliminary Surveys

Preliminary surveys are data-gathering surveys used to determine the specific locations of existing natural and man-made features at the job site. The differences in elevation of various points at the site are also determined. This information is subsequently plotted to scale on a map or site plan, including the information about the site elevations and contours.

3.2.3 Construction Surveys

Construction surveys (*Figure 2*), also called layout surveys, are performed at the job site on a continuous basis to locate structures and provide required elevation points in accordance with the project design plans. During these surveys, the horizontal location and the vertical location or elevation for each structure or feature is laid out.

101F02.EPS

Figure 2 ◆ Surveying on a construction site.

3.2.4 As-Built Surveys

As-built surveys are performed to obtain information about a project or a portion of a project after the project has been completed. The survey ties in all the job-site features to provide a final record of what was constructed and to verify that the construction

was done in accordance with the plans and specifications. Information derived from the as-built survey is usually recorded in the form of as-built plans or maps.

3.3.0 Specific Types of Surveys

In addition to the types of surveys previously described, other types of surveys have the following specific applications:

- Property surveys
- Route surveys
- Topographic surveys
- Hydrographic surveys

3.3.1 Property Surveys

Property surveys are a type of plane survey. They are done to locate property lines and corners, to determine the boundaries pertaining to parcels of land, and to plan and lay out subdivisions of property. Property surveys are normally required by law whenever real estate transactions take place.

Surveyors, field engineers, and trainees under the supervision of surveyors and field engineers can perform most surveying tasks at a job site. However, because of the tremendous liability involved, most states require that only a registered professional surveyor perform any surveying that establishes legal property lines and boundaries. A specific knowledge of boundary law is required for decisions involving deed interpretation and boundary determination.

3.3.2 Route Surveys

Route surveys pertain to surveys of long and narrow areas through which highway, railroad, canal, gas or power line, or similar paths are routed. The survey involves the determination of the **relief** and location of natural and man-made objects along the proposed path or route. Route surveys often involve staking out the proposed route and calculating earthwork quantities.

3.3.3 Topographic Surveys

Topographic surveys record both the horizontal plane measurements of the area being surveyed and the changes in elevation. They are performed to gather detailed information about the locations and elevations of all natural and man-made features within the surveyed area. Topographic maps and survey drawings (*Figure 3*) based on the survey data include lines called **contour lines** that represent the various elevations within the surveyed area. A topographic survey may be either a plane survey or geodetic survey, as determined by the size of the survey area involved.

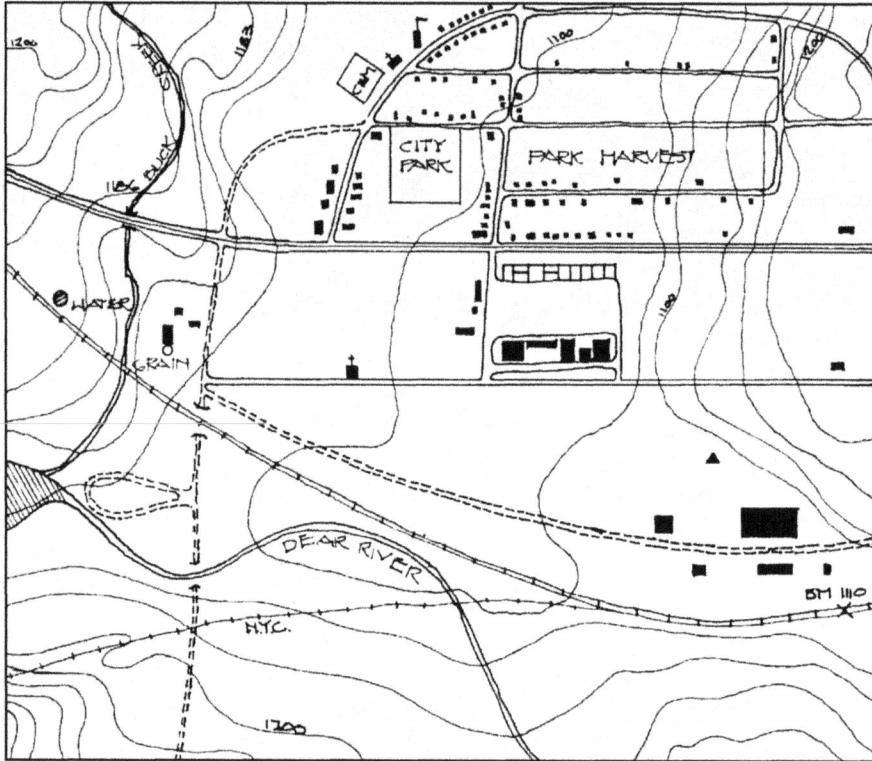

TOPOGRAPHIC MAP

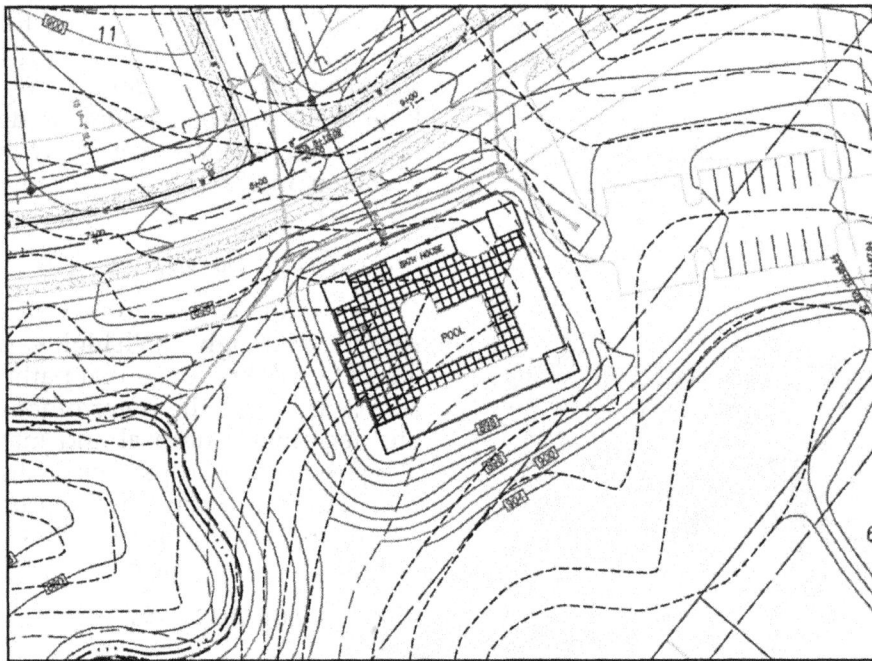

TOPOGRAPHIC SURVEY DRAWING   101F03.EPS

Figure 3 ◆ Examples of a topographic map and a topographic survey drawing.

3.3.4 Hydrographic Surveys

Hydrographic surveys are similar to topographic surveys except that the media being surveyed involves bodies of water instead of land. Hydrographic surveys typically are made for rivers and lakes to determine their shoreline, underwater features, and flow rate for applications such as navigation and flood control. Other hydrographic surveys, called marine surveys, cover surveying of larger bodies of water. Marine surveys are typically associated with surveying for offshore platforms and the preparation of hydrographic charts and maps.

4.0.0 ◆ LATITUDE AND LONGITUDE AS GEOGRAPHIC REFERENCES

The most basic geographic reference system used to pinpoint the location of physical land features on Earth's surface consists of imaginary geographic lines called latitude and longitude. As shown in *Figure 4*, latitude lines run horizontally east-west on Earth's surface. They run parallel to the equator and to each other, and they are of equal distance from each other. The equator is an imaginary 0° line that divides Earth into the northern and southern hemispheres. Starting at the equator, latitude lines are identified from 0° to 90° north (+ latitudes) and 0° to 90° south (– latitudes). Each degree of latitude is approximately 69 miles apart. There is a slight variation in the distances between latitudes because Earth is slightly egg-shaped (elliptical) instead of being perfectly round.

Longitude lines, also called meridians, run vertically north and south and pass through different parts of Earth's surface before converging at the north and south poles. They are at their widest (about 69 miles apart) at the equator. However, the lines of longitude get closer together as they approach the poles. A prime meridian located at Greenwich, England is designated as 0° longitude. This meridian divides Earth into the Eastern and Western hemispheres. Starting from the 0° longitude meridian, longitude lines extend around Earth 180° to the east (+ longitudes) and 180° to the west (– longitudes) until they meet in the Pacific Ocean. The point in the Pacific Ocean where they meet is called the International Date Line.

As you can see, Earth's latitude and longitude lines form a grid where latitude is a measure of how far north or south a place is from the equator. Similarly, longitude is a measure of how far east or west a place is from the prime meridian. To enable points on Earth to be more accurately located, each degree of latitude and longitude is divided further into minutes and seconds, with each degree being equal to 60 minutes (60'), and each minute being equal to 60 seconds (60"). In terms of distance, this means at the equator one minute is equal to about 1.15 miles (69 miles ÷ 60 minutes) and one second is equal to about 101 feet [(1.15 × 5280) ÷ 60].

Some examples of locations stated in terms of latitude and longitude are New Orleans, Louisiana, located at 29° 59' N latitude and 90° 15' W longitude; and Washington, DC, located at 38° 53' 23" N latitude 77° 00' 27" W longitude. Locations identified by their latitude and longitude coordinates are also commonly expressed in term of plus (+) and minus (–) degrees of latitude and longitude. For example, the New Orleans location can also be expressed as +29° 59' latitude, –90° 15' longitude.

Latitude and longitude coordinates are used mainly for navigation and some special types of surveying. Surveys performed for site layout are typically plane surveys, the points of which are referenced to state and/or local control grid monuments.

5.0.0 ◆ SITE LAYOUT CONTROL POINTS

Site layout involves establishing a network of control points that serve as a common reference for all construction on a site. The exact locations of these control points are marked at the site and recorded in the field notes as they are made. Annotating control point location reference data in the field notes is important for two reasons. First, it makes it possible to locate a point should it become covered up or otherwise hidden. Second, it makes it possible to reestablish a point accurately if the marker is damaged or removed.

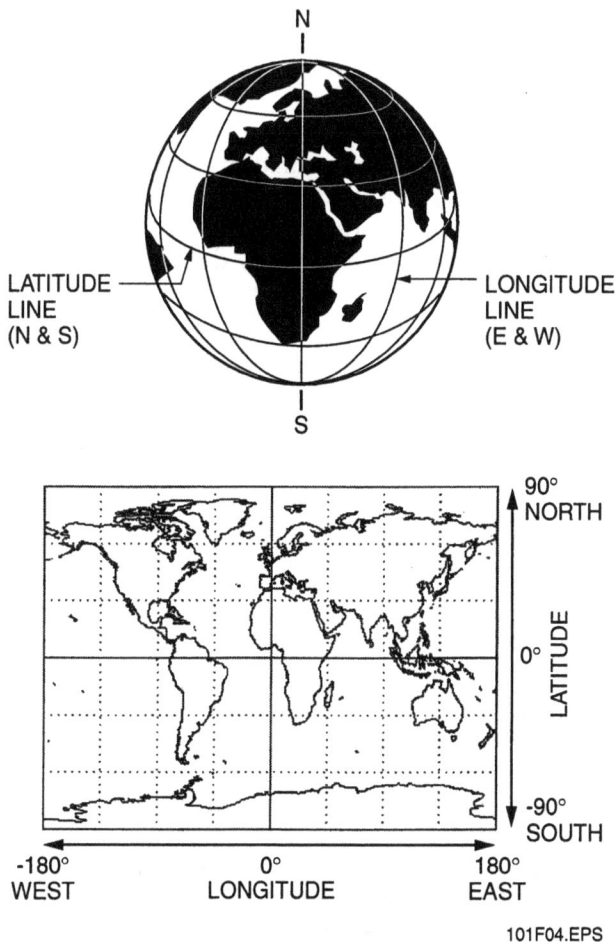

101F04.EPS

Figure 4 ◆ Earth's latitude and longitude lines.

The are four categories of control points are listed below in the order of their sequence and required accuracies. Collectively, these control points form a complete control network used for performing various site layout tasks.

- Reference azimuth control points
- Primary control points
- Secondary control points
- Building layout or working control points

In most states, registered surveyors are required to perform any layout work that establishes legal property lines or boundaries. This is because the surveyor assumes the liability for any mistake in the surveying work. The surveyor is legally responsible if the building ends up on the wrong property or at some location that violates setback requirements or other regulations. Because of the liability involved, the site layout technician should never make any layout of legal property lines or boundaries. This is a task for the professional land surveyor.

5.1.0 Reference Azimuth Control Points

Reference azimuth points (RAPs), also called main reference points, are one or more control points established either on or off the project site. Permanent RAPs are usually located and marked by architectural or engineering firms during the pre-job planning and layout phases of a project. RAPs are easily identifiable and visible points such as communications towers, church spires, or similar structures. The coordinates describing the location of RAPs are typically determined by measuring angles to the RAPs from previously located state or local control grid monuments. RAP locations are used as reference control points from which a grid of primary control points is established on the job site.

5.2.0 Primary Control Points

Primary control points are permanent points used as the basis for locating a grid of secondary control points on the job site. Primary control points are located where they are accessible and protected from damage for the duration of the job. Primary control points can be located and marked on many kinds of permanent and immovable objects such as fire hydrants and power poles. When no suitable permanent objects are available for use as a primary control point marker, man-made markers can be used, such as iron stakes driven into the ground or a concrete monument dug and poured into the ground (*Figure 5*). If a poured concrete monument is used, it must be dug a foot deeper than the frost line to prevent

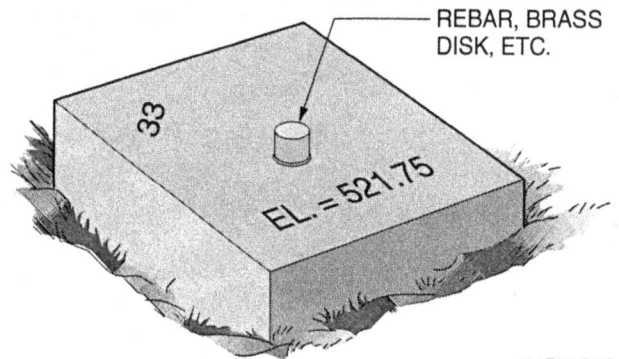

Figure 5 ◆ Control point concrete monument.

frost heave. It must also have a distinct high point, such as a rounded brass cap or rebar, that sticks up out of the top of the concrete.

Primary control point markers are typically established by a registered surveyor. They are also called monuments or bench marks.

5.3.0 Secondary Control Points

Secondary control points are additional control points located within the job site to aid in the construction of the individual structures on the site. These points are usually established by the contractor, but may be done by the surveyor in some cases. Secondary control points are typically marked by a hub stake surrounded by protective laths (*Figure 6*), posts, or fencing. The hub stake is typically a 1½" square piece of wood pointed on one end. Its length is normally determined by the hardness of the ground it must be driven into, with lengths between 8" and 12" being typical. The hub stake is driven into the ground until flush or nearly so. A surveyor's tack, with a depression in the center of the head, is driven into the top of the hub stake to locate the exact point.

5.4.0 Building Layout or Working Control Points

Building layout or working control points are usually located with reference to the secondary control points. These are the points from which actual measurements for construction are taken. Building layout points are used to locate the corners of buildings and building lines. They are usually marked with a hub and a related marker stake (*Figure 7*). The marker stake is typically a ¾" × 1½" piece of wood that varies in length, with 24" to 36" being typical.

In addition to serving as hub markers, these stakes are used to mark line or grade and other information, such as center lines, offset lines, and slope.

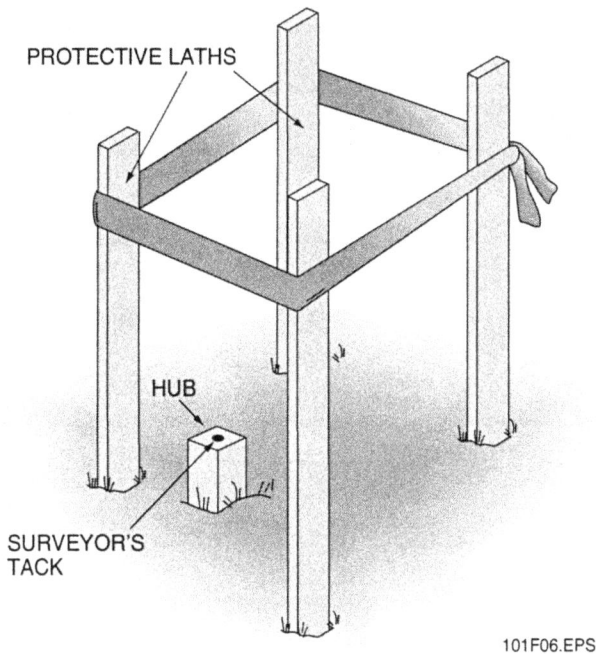

Figure 6 ◆ Secondary control point marker.

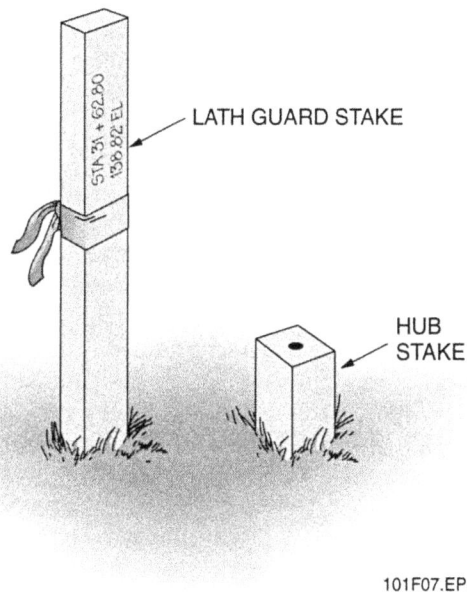

Figure 7 ◆ Working control point marker.

5.5.0 Placement of Control Points and Other Markers

The placement of the numerous on-site control point markers depends on the nature of the job, the terrain, other work in progress, the sequence of work, and many other factors. Accuracy in site layout work requires that bench marks, control points, and other important markers be referenced in a way that ensures they can be easily located at a later date. Good practice dictates that a control point should always be placed so that three other control points are visible at all times.

This is done in case the control point should become covered up or destroyed for some reason. If this happens, two distance measurements can be made from the other points in order to relocate the control point. These points should be referenced to permanent objects in the surrounding area. Guidelines for establishing such references are summarized here.

- Establish several (three or more) definable permanent or semi-permanent references for each point:
  - A bonnet bolt located on a nearby fire hydrant
  - Wooden power poles
  - Sidewalks
  - Trees
  - Fences
  - Building corners
  - Sign posts
- To locate a point more accurately, reference it in all directions (north, south, east, west) from the point rather than in the same direction, so that when arcs are swung from the reference points, there is a very small, distinct area where they intercept.
- If possible, stay within one tape length of the point you are using as a reference.
- Draw a clear, complete sketch in the field notes.

The placement and positioning of stakes at the construction site must be done properly. Some guidelines for performing this task include the following:

- Face the stakes so that they can be read from the direction of use.
- Offset the stakes as required for their protection.
- Set the stakes within tolerances.
- Place the stakes solidly in the ground.
- Place the stakes and laths so that they are plumb.
- Center the hubs and stakes.

5.6.0 Communicating Information on Control Markers and Other Markers

Hubs, stakes, and laths must be legibly and accurately marked to correctly communicate location, elevation, and other pertinent construction information (*Figure 7*). Some guidelines for marking stakes are as follows:

- Use a permanent marker.
- Print neatly. Start from the top of the stake and work toward the bottom. Use all capital letters. Be careful not to crowd your words or numbers.

- Avoid the use of abbreviations or use standard abbreviations.
- All sides of a stake can be marked with information; however, the main information should always be marked on the direction of use.
- Mark only pertinent information. Too much or too little information can be confusing.
- Mark the following types of information on a stake, as applicable:
    - Alignment information
    - Center line
    - Cut or fill data
    - Elevation data
    - General information description
    - Grade information
    - Offset
    - Reference information
    - Slope
    - Specific information

In addition to marking stakes, it is often necessary to mark reference lines and points on walls, curbs, foundations, and other areas. When doing so, the same general guidelines should be followed that were given for marking information on stakes. In addition, any such reference lines must be marked straight, level, and/or plumb. *Figure 8* shows an example of a reference line marked on a wall 4' above a finished floor.

101F08.EPS

Figure 8 ◆ Example of a point marked 4' above the finished floor.

5.7.0 Color-Coding Control Markers

Control points and other markers are identified by color-coding both to identify their purpose and so that they can be easily seen and recognized. Color-coding can be done by applying paint to monuments, hubs, or other field markers; applying ribbons on stakes; or attaching flags on wire markers. Note that field marker color codes are not standardized; however, many construction and field engineering organizations have established their own color-coding systems. When performing layout work at a site, you should ask your supervisor what color-coding scheme to use and follow it.

> **NOTE**
> In some states, mandatory color coding schemes are used. A contractor can be fined for failing to use the correct color.

6.0.0 ◆ MEASUREMENTS

Numerous kinds of measurements must be made when performing site layout work, including measurement of distances, elevations, and directions. The following factors pertain to all measurements, regardless of the kind of measurement.

6.1.0 Accuracy and Precision

Accuracy is defined as how close a measurement of a distance, angle, or elevation being made is to its true value. The true value is defined as the mean value as determined mathematically by a series of repeated measurements. Precision is defined as the degree of refinement with which a measurement is made. It is a comparison of the closeness of a given measurement to another measurement of the same quantity. The goal in site layout surveying is to make all your measurements both accurate and precise. Precise measurements based on faulty datum are precisely wrong.

Figure 9 uses a series of archery targets to help explain the concepts of accuracy and precision. As shown in view A, the grouping of the arrows is precise (close together), but not accurate relative to the position of the bull's eye. In view B, the grouping of the arrows is more accurate relative to the position of the bull's eye, but not precise. In view C, the grouping of the arrows is both accurate and precise relative to the bull's eye.

6.2.0 Errors and Mistakes

Errors exist in every measurement. Error is defined as the difference between a measured quantity and its true value. Errors result from both human and instrumentation limitations and from the effects of weather. Errors are classified as being either **random errors** or **systematic errors**. Human errors are considered to be random errors. The characteristic of random errors is such that their magnitude is unknown and that with repeated measurements their effect tends to cancel out.

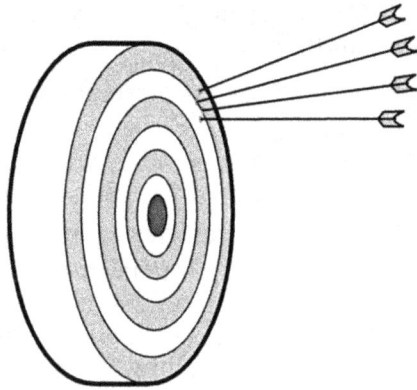

(A) PRECISE BUT NOT ACCURATE

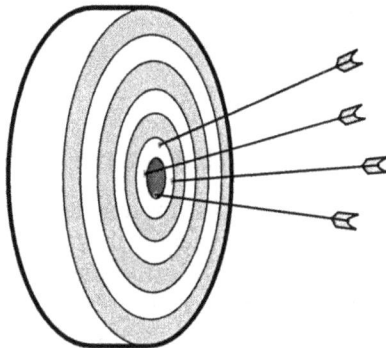

(B) ACCURATE BUT NOT PRECISE

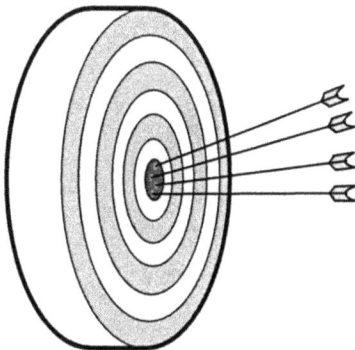

(C) BOTH ACCURATE AND PRECISE

101F09.EPS

Figure 9 ◆ Accuracy and precision.

Instrumentation errors are considered to be systematic errors because they occur repeatedly in the same magnitude when making measurements. Instrumentation errors can occur because of instruments or equipment being damaged, out of calibration, or incapable of measuring within tolerances. These types of errors are compensated for by making sure you always use well-maintained and calibrated measuring instruments. An example of a systematic error is a steel measuring tape that has an actual length different from its marked length being used to make distance measurements. This can result from expansion or contraction of the tape caused by ambient temperature variations. This type of systematic error is compensated for by making corrections to the measured values via a mathematical formula.

Mistakes result from carelessness or a misunderstanding when making and/or recording measurements. Mistakes can be eliminated if an organized approach is used and the proper measurement procedures are followed. Mistakes will happen, but they can be discovered and corrected by checking, then rechecking your work.

6.3.0 Accuracy Ratio

The accuracy ratio for a measurement or series of measurements is the ratio of error of closure to the distance measured. The error of closure is the difference between the measured location and the actual or correct location. The accuracy ratio is defined as the error in a measurement divided by the overall value of the measurement. It is expressed as a fraction with a numerator of 1 and a denominator rounded to the closest 100 units. For example, a distance was measured and found to be 196.33'. The distance was previously determined to be 196.28'. The error is 0.05' in a distance of 196.28'. This error results in an accuracy ratio of $\frac{1}{3900}$ derived as follows:

Accuracy ratio = 0.05/196.28 = 0.000254738

= 1/0.000254738 = 3,925.6

= 3,900 (rounded to nearest 100 units)

6.4.0 Tolerances

Tolerances are used in construction layout to define how far off the exact design location something can be and still be acceptable. You must always meet or exceed the specified tolerances. Several factors determine the degree of accuracy required for a specific job or different control points within a job site. Once the specific tolerances are known, you must use measurement equipment capable of measuring to the required tolerances.

7.0.0 ◆ DISTANCE MEASUREMENT TOOLS AND EQUIPMENT

Site layout often involves making horizontal and vertical distance measurements by a process called taping or chaining. The term chaining is derived from past surveying practices when a metal chain was used to measure distances. Today, the terms taping and chaining are used interchangeably in the trade when referring to the measurement of distances using a steel tape or other type of tape. In this module, we will use the term taping when referring to the distance measurement process and the term tape (instead of chain) when referring to the measuring tape. The task of taping involves two people working together and communicating with each other. The major items of equipment (*Figure 10*) needed to perform taping typically include the following:

- 100' or longer steel tape
- Plumb bobs and gammon reels
- Hand sight levels
- Accessories and stakes

7.1.0 Tapes

Tapes can be made of cloth, fiberglass, or steel. However, steel tapes are the most widely used for precision measuring tasks. Steel tapes are made in a variety of graduated lengths, with 100' being common. Typically, the tape is mounted on a reel for ease of handling and storage. Tapes are made with graduations in feet and inches or feet and tenths of a foot. Metric versions are also available. Some steel tapes are nylon coated to increase their durability. The ends of steel tapes are equipped with heavy loops that provide a place to attach leather thongs or tension handles. These allow the user to tighten or tension the tape firmly. The physical construction of a tape requires that it be handled and used properly.

These are some guidelines for the proper care and handling of tapes:

- Keep the tape on the reel and rolled up when not in use.
- Dry a tape that is wet. Once dry, wipe it with an oiled cloth for added protection against rust.
- Do not allow vehicles to run over the tape.
- When making measurements, the proper amount of tension must be applied to get accurate results. The tension is specified by the tape manufacturer. Pulling too hard will eventually stretch and permanently elongate the tape. If a tension spring is not available, try to pull as consistently as possible.
- Remove all loops in a tape immediately. This is important to prevent kinks from deforming the tape.

7.2.0 Plumb Bobs and Gammon Reels

When suspended from a string, a plumb bob provides a plumb vertical line that you can use to position yourself or an instrument directly over a reference point. Plumb bobs are usually made of brass and have a replaceable tip. Depending on the model, they can weigh between 8 and 24 ounces, with 16 ounces being typical. The gammon reel is used to store the plumb bob string. The string

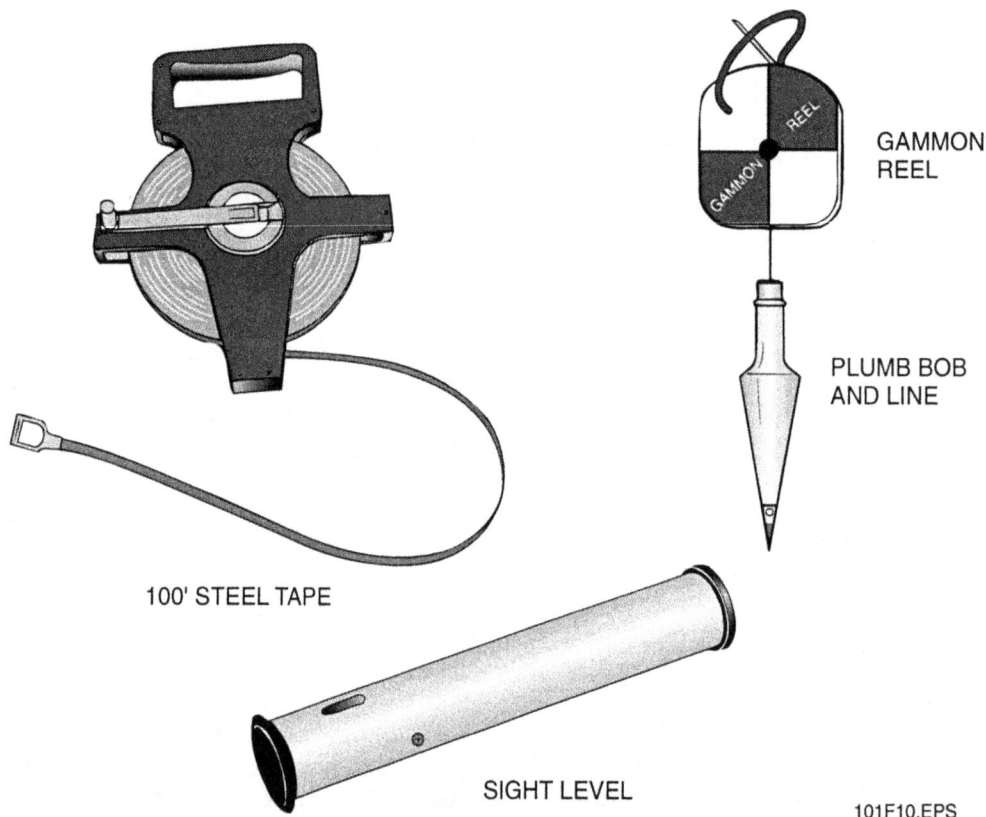

GAMMON REEL

PLUMB BOB AND LINE

100' STEEL TAPE

SIGHT LEVEL

101F10.EPS

Figure 10 ◆ Common taping equipment.

automatically retracts into the reel to help prevent the string from becoming tangled, broken, or muddy. The case of the gammon reel is colored so that it can be used as a target for sighting purposes. These are some guidelines for the proper care and handling of plumb bobs and gammon reels:

- Keep plumb bobs and gammon reels clean.
- Never use a plumb bob as a hammer.
- Do not use a plumb bob tip to mark hard surfaces because this will damage the tip.
- Clean the plumb bob string before allowing it to be retracted into the gammon reel. This prevents dirt from getting inside the gammon reel, where it can damage the retracting mechanism.
- Check the string for wear and knots and replace when necessary.

7.3.0 Hand Sight Levels

When you are taping distances or making layout measurements, a hand sight level can be used to determine your position on line and/or the correct horizontal position of the tape needed to plumb measurement points. The sight level is a short, handheld telescope with a bubble level built into it. The bubble level is visible when sighting through the telescope. When the bubble is centered, the **crosshairs** of the scope fall on some object that is at about the same elevation as the eye of the user.

For leveling tasks, a hand level can be used to help determine where to set up the leveling instrument so that its line of sight will intercept the leveling rod. This is done by using a hand level to sight on a bench mark or other object to see if the height is above or below eye level. These are some guidelines for the proper care and handling of a sight level:

- Do not drop the level.
- Wipe the lens with a clean cloth as needed.
- Keep the level in its protective case when it is not being used.
- Check its calibration frequently.

7.4.0 Other Equipment

Other types of equipment that may be used in conjunction with the taping process include a tension spring (*Figure 11*), a thermometer, and clamps. A tension spring is a spring scale device that can be attached to the end of the tape for the purpose of measuring the amount of pull or tension that is applied to the tape. Its use takes the guesswork out of how much tension is being applied to the tape during a measurement. Its

repeated use over successive measurements helps yield more accurate measurements by allowing the user to apply the same amount of tension to the tape each time it is used. Clamps are used to grip a tape anywhere along its length to protect both the user's hands from cuts and the tape from bending or breaking. Thermometers are used to take temperature readings of the tape when it is necessary to correct the values measured with a tape to compensate for temperature variations.

101F11.EPS

Figure 11 ◆ Spring-balanced tension spring.

8.0.0 ◆ MEASURING DISTANCES BY TAPING

To achieve both accuracy and precision when taping, you must have a good understanding of the measurement process and its principles.

Guidelines that must be followed in order to achieve accuracy when taping are summarized in *Figure 12* and briefly explained in this section.

- Use a calibrated tape. Manufacturers furnish a data sheet with each precision tape they make. This sheet will tell you exactly how accurate the tape is and to which standard it is compared.
- Determine where the exact zero point of the tape is, and use it. It may be at the end of a loop or other fitting at the end of the tape, or it can be offset from the end of the tape.
- Maintain good alignment. Distances must be measured as a straight line. When measuring distances that are longer than the tape, intermediate measurement points must be used. Any such intermediate points should be directly in line between the beginning and end points being measured.
- Apply the correct tension to the tape during measurements. This is necessary to overcome any sag in the tape. For the correct amount of tension to use with a specific tape, use the tension value specified by the manufacturer in the product literature supplied with the tape. As mentioned earlier, a tension spring can be attached to the end of the tape to aid you in applying the right tension.

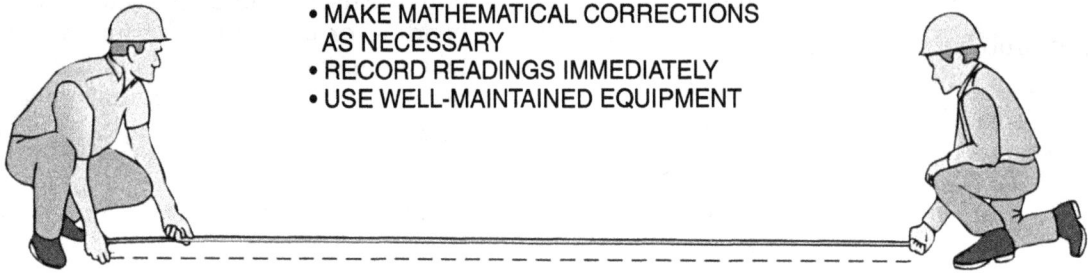

- USE A CALIBRATED TAPE
- KNOW WHERE ZERO IS ON THE TAPE
- MAINTAIN GOOD ALIGNMENT
- APPLY PROPER TENSION TO THE TAPE
- MEASURE HORIZONTALLY
- REPEAT MEASUREMENTS
- MAKE MATHEMATICAL CORRECTIONS
  AS NECESSARY
- RECORD READINGS IMMEDIATELY
- USE WELL-MAINTAINED EQUIPMENT

101F12.EPS

Figure 12 ◆ Guidelines for distance measurements by taping.

- The tape must be read when it is in a horizontal position. On flat ground, this is not much of a problem. On a slope or incline, tape readings can be taken in smaller increments. This allows the downhill person to comfortably hold the tape in a horizontal position while still applying the right tension to the tape. Typically, this puts the end of the tape about chest high.

- To avoid mistakes that can easily occur when taping, make all measurements at least twice. Reversing the direction of the measurements greatly reduces the chance of repeating a mistake.

- Make mathematical corrections as necessary. Mathematical corrections must be made to measurement data to compensate for conditions such as:

  - Tape length differences resulting from calibration
  - The expansion or contraction of the tape's length due to temperature variations
  - Sloped (rather than horizontal) placement of the tape during a measurement

  All of these conditions can affect the accuracy of your measurements. These conditions are described in more detail later in this module.

- Record the reading for each measurement in the field notes immediately after it is taken in order to avoid omissions that can contribute to errors. Also, be sure to check that you have recorded each entry correctly. It is easy to transpose numbers or misplace a decimal point.

- Use well-maintained equipment. Accurate measurements can only be made with well-maintained taping equipment.

8.1.0 Taping a Distance

The task of taping involves two people working together and communicating with each other. The following procedure outlines one method for measuring a distance between two existing points, such as two control monuments. The procedure for measuring a distance between known and unknown points, such as when laying out a building, is performed in basically the same manner. In this procedure, the two people involved are designated as the rear tape person and the head tape person (*Figure 13*). For the purpose of explanation, it is assumed that a 100' tape is being used, the overall distance to be measured is greater than 100', and the terrain is relatively flat, allowing for horizontal measurements to be made with the tape on the ground.

> **NOTE**
>
> Different individuals, including surveyors, field engineers, and trainees under the supervision of surveyors and field engineers, can perform most site preparation and layout tasks. However, because of the tremendous liability involved, most states require registered professional surveyors to perform any layout work that establishes legal property lines and boundaries. This is because the surveyor assumes the liability for any mistakes in the surveying work.

Step 1 Determine the straight path for the overall measurement between the start and end points of the line to be measured. This path should be cleared of any brush, rocks, or other obstacles that will hamper making the measurements.

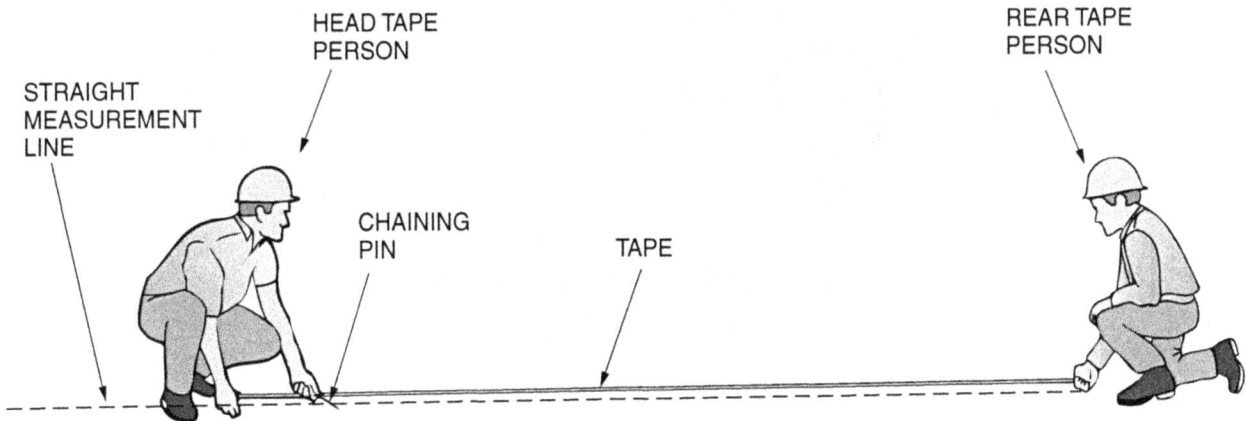

Figure 13 ◆ Taping procedure.

Step 2 Once the measurement path has been determined, the location of the start and end points of the overall measurement path are marked with laths with flagging or other suitable markers.

Step 3 While the rear tape person holds onto the 0' end of the tape at the beginning point, the head tape person takes the tape reel and advances along the measurement line to the location of the first intermediate measuring point, presumably at 100' from the beginning point. Some companies prefer that the head tape person advance with the 0' end of the tape while the rear tape person holds onto the reel. Either way is acceptable.

Step 4 The head tape person applies the proper tension to the tape as the rear tape person aligns the 100' mark on the tape exactly on the starting point. Once positioned over the starting point, the rear tape person signals the head tape person, who then marks the position of the 0' point of the tape with a suitable marker, such as a chaining pin.

Step 5 Following this, both workers advance along the measurement line in preparation for the next measurement. The starting point for this measurement is at the location of the chaining pin established in Step 4.

Step 6 Steps 3 and 4 are repeated as required until the line has been measured from the original starting point to the end. To avoid mistakes, all distances should be measured at least twice.

Step 7 The overall distance measured is equal to the total number of full tape lengths measured plus the reading of the last measurement. For example, if three 100' tape measurements were made, and the last measurement recorded by the rear tape person is 30.25'; then the total distance measured is 330.25'.

The general procedure for measuring a distance over terrain with an excessive slope or incline is basically the same as described above. However, to aid in maintaining the tape in the horizontal (level) position when making measurements, a method called **breaking the tape** is used. Simply, this means making measurements using a portion of the full tape's length in a series of steps until the full tape length has been **traversed**.

For example, if moving down the slope of the hill, the head tape person advances with the end of the tape along the line for a distance equal to the tape length. Then, leaving the tape on the ground, the head tape person returns as far along the tape as necessary to reach a point that will allow him or her to comfortably hold the tape in the horizontal position and still be able to apply the proper tension during the measurement (*Figure 14*). Typically, this is about chest or waist high.

8.2.0 Corrections for Tape Temperature

Tape manufacturers normally calibrate steel tapes when the tapes are at a standard temperature of 68°F. Steel tapes expand with increasing temperatures and contract with decreasing temperatures. Steel tapes have a thermal expansion coefficient of 0.0000065 feet per foot per 1°F. This means that for every 1°F change in temperature, the length of the tape will change by 0.0000065' from its original

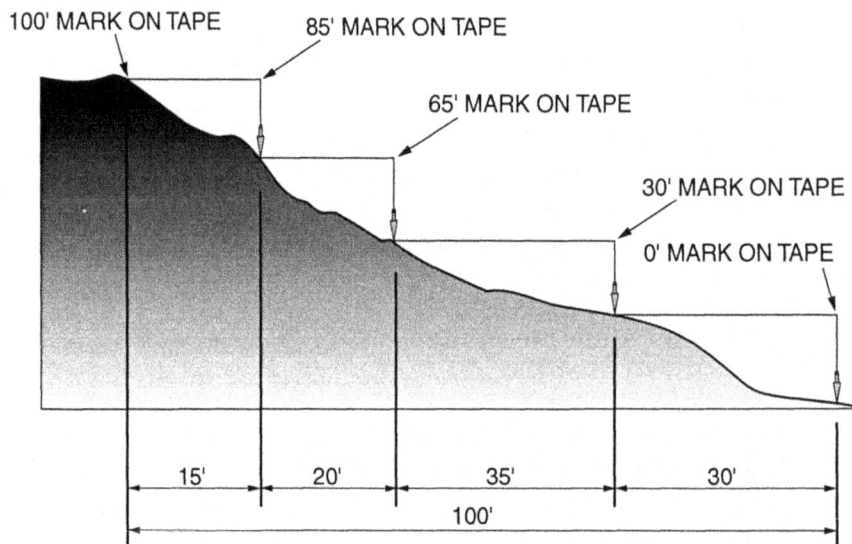

Figure 14 ◆ Breaking the tape to measure a distance on a steep slope.

length. To compensate for this error, the temperature of the tape at the time of measurement must be known. This is typically done by measuring the temperature of the tape with a thermometer, then calculating and applying any correction to the measurement. The temperature should be measured at the tape; the temperature of a tape lying on a hot surface, such as an asphalt road, or on a cold surface, such as the ground, can be considerably different from the temperature of the surrounding air.

To compensate for tape length changes caused by temperature, the total distance measured in feet is multiplied by the coefficient of expansion (0.0000065). Then the result is multiplied by the value of the temperature difference between the actual tape temperature and the standard temperature:

$$\text{Correction } (C_t) = D \times 0.0000065 \times (T_t - 68)$$

Where:

D = recorded distance or distance to be laid out

T_t = temperature of the tape at the time of measurement

When the temperature of a tape is warmer than the standard temperature of 68°F, the tape expands, making it too long relative to its calibrated length. Conversely, when it is cooler than the standard temperature, the tape contracts, making it too short.

9.0.0 ◆ SITE LAYOUT INSTRUMENTS AND EQUIPMENT

A wide variety of optical and electronic instruments are used to perform distance measurements, leveling, and angular measurement site layout tasks. The builder's level and related equipment are covered here.

9.1.0 Builder's Level

The builder's level (*Figure 15*) is an instrument used to check and establish grades and elevations, to set up level points over long distances, and to turn horizontal angles when laying out building lines. A builder's level cannot be used to measure vertical angles. It consists of a telescope, a bubble spirit level (leveling vial) mounted parallel with the telescope, and a leveling head mounted on a circular base with a horizontal circle scale graduated in degrees. The telescope can be rotated 360° for measuring horizontal angles and can be tilted slightly for sighting purposes. Builder's levels are mounted on a tripod when in use.

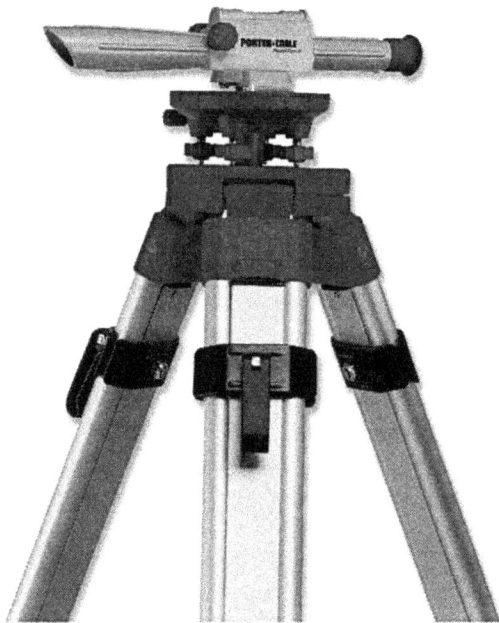

101F15.EPS

Figure 15 ◆ Builder's level.

Depending on the model, builder's levels are made with telescope powers ranging from 12 power (12X) to 32 power (32X), with 20 power (20X) being the most common. The power of a telescope determines how much closer an object will appear when viewed through the telescope. Two types of leveling head systems are used in builder's levels: a four-screw system and a three-screw system. The advantage of the three-screw system is that it allows the instrument to be leveled

more quickly. Four-screw systems are more common in older models. Coverage of the builder's level in this module focuses on its use in determining elevations.

9.2.0 Tripods

Because most surveying activities are performed in the field or out on the job site, it is not always easy to get sighting and leveling devices properly aligned to a given reference point. The environments in which these reference points exist may damage the surveying equipment if it is placed onto the ground or floor. Placing the equipment on the ground or floor may also make it very difficult for the instrument person to see through the sighting device; therefore, a tripod is used.

9.2.1 Tripod Basics

Many years ago, devices called tripods were developed so that the reference point could be offset vertically and the instruments could be better positioned and more easily used. Most tripods are made of wood, aluminum, or fiberglass, and, as the name implies, they have three supporting legs. Each leg is adjustable so that the tripod can be set up on almost any surface. *Figure 16* shows a typical tripod that can be used with most surveying instruments.

9.2.2 Tripod Mounting Plate and Leveling

The mounting plate on top of a tripod is typically flat and may or may not have small levels attached to it. Some mounting plates or heads have mounting rings on top of the plates for securing the instruments to the plates. Some of these mounting rings have threads inside them, and others have mounting threads outside the ring. Some form of hole in the center of the mounting plate allows for a plumb bob to be suspended from the instrument mounted on the tripod. *Figure 17* shows tripod mounting plates with instrument mounting rings for external or internal threading.

If the mounting plate of a tripod does not have levels attached to it already, a small construction level can be laid across the top of the plate to check the level of the plate. The legs of a tripod must be adjusted so that the mounting plate is level in all horizontal directions. Ensuring that the tripod is properly leveled and securely set is critical for good readings from the instruments attached to the tripod. Making sure that the tripod legs are securely set also helps ensure that the tripod does not get easily knocked over and possibly damage some very expensive instrument attached to it.

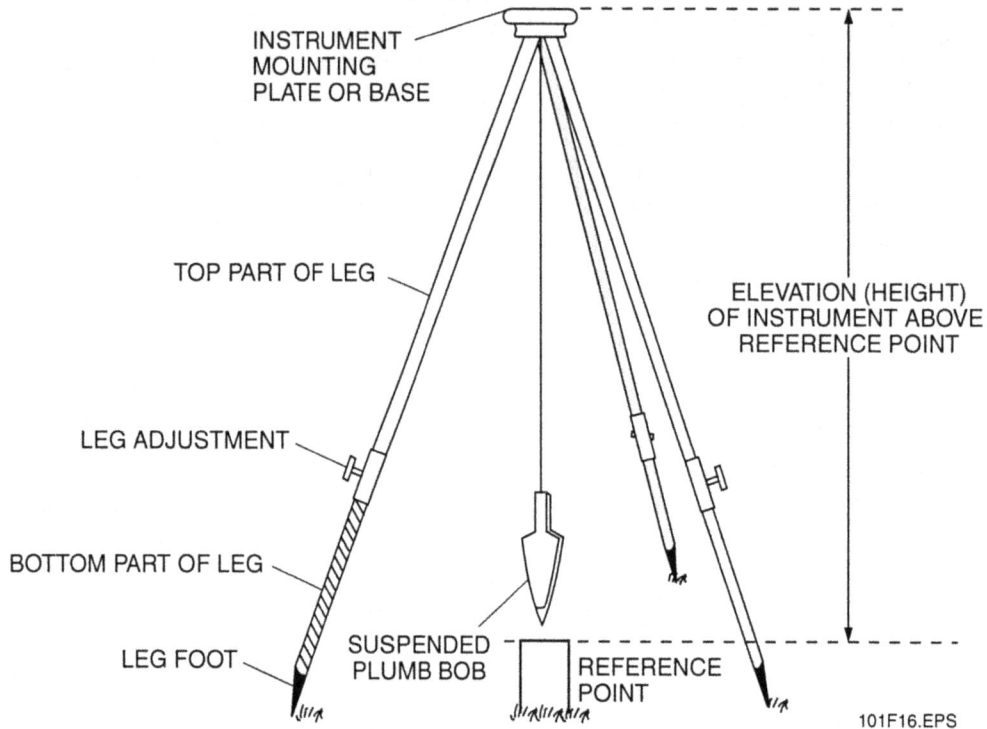

Figure 16 ◆ Surveying tripod.

Labels in figure:
- INSTRUMENT MOUNTING PLATE OR BASE
- TOP PART OF LEG
- LEG ADJUSTMENT
- BOTTOM PART OF LEG
- LEG FOOT
- SUSPENDED PLUMB BOB
- REFERENCE POINT
- ELEVATION (HEIGHT) OF INSTRUMENT ABOVE REFERENCE POINT
- 101F16.EPS

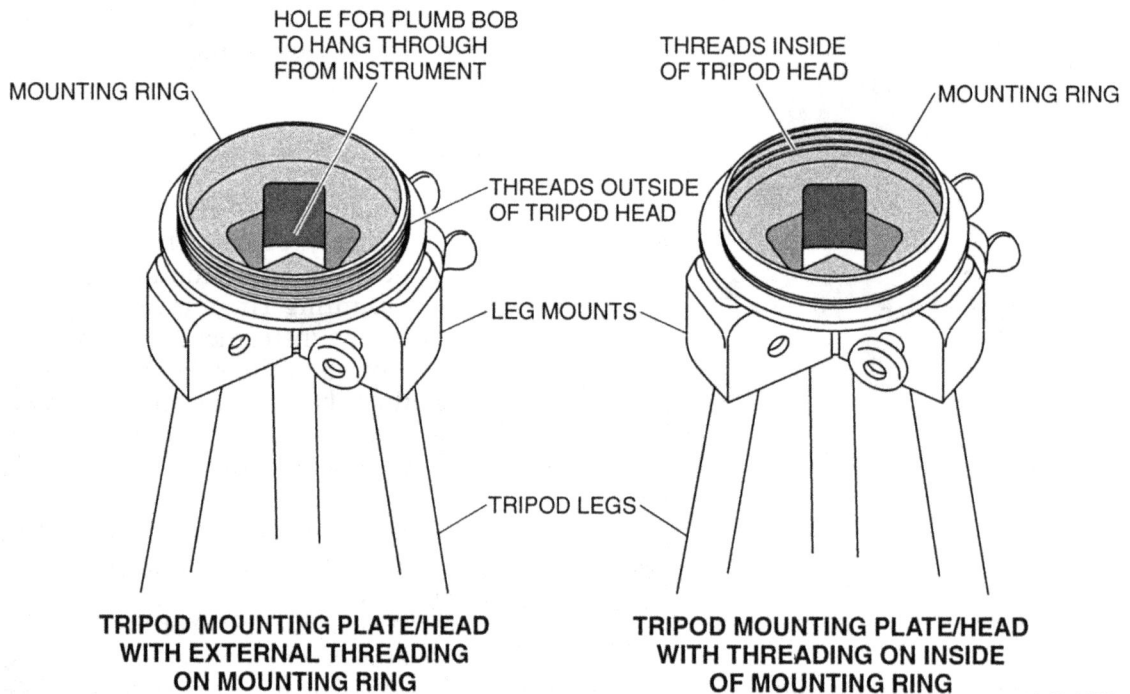

Labels in figure:
- HOLE FOR PLUMB BOB TO HANG THROUGH FROM INSTRUMENT
- MOUNTING RING
- THREADS INSIDE OF TRIPOD HEAD
- MOUNTING RING
- THREADS OUTSIDE OF TRIPOD HEAD
- LEG MOUNTS
- TRIPOD LEGS

TRIPOD MOUNTING PLATE/HEAD WITH EXTERNAL THREADING ON MOUNTING RING

TRIPOD MOUNTING PLATE/HEAD WITH THREADING ON INSIDE OF MOUNTING RING

101F17.EPS

Figure 17 ◆ Variations of tripod mounting plates and rings.

9.2.3 Tripod Maintenance

When the tripod is no longer needed, it needs to be removed from the work location, cleaned, lubricated as needed, inspected for damage, and properly stored if it is undamaged. Surveying instruments are only as good as the foundation on which the instruments are mounted. Taking care of the tripod on which the instruments sit is very important. Guidelines for the use and proper care and handling of tripods are as follows:

- When setting up the tripod, position the tripod legs properly. The legs should have about a 3' spread, positioned so that the top of the tripod head is horizontal.
- If the tripod's legs are adjustable, make sure that the leg levers are securely tightened.
- If setting up on dirt, make sure that the tripod points are well into the ground. Apply your full weight to each leg to prevent settlement.
- When setting up on a smooth floor or paved surface, secure the points of the legs by attaching chains between the legs or putting a brick or similar object in front of each leg.
- Attach the instrument to the tripod securely. Do not overtighten the attaching hardware.
- Frequently lubricate the joints and adjustable legs of the tripod using an appropriate lubricant.
- When not in use, protect the head of the tripod from damage.
- When transporting a tripod in a vehicle, never pile other materials on top of the tripod. Make sure to protect it from damage that can be caused by shifting equipment or materials.
- Keep the tripod clean and dry. When not in use, store the tripod in its protective case.

9.2.4 Tripod Setup

The first step in preparing to set up a tripod to be used with an instrument is determining which type of instrument is to be mounted on the tripod. Another consideration is ensuring that there is enough working space around the base of the tripod. The person using the instrument mounted on the tripod will need room to walk around the tripod legs. Perform the following steps to set up a tripod:

Step 1 Determine the following before selecting a tripod for the job:
- The kind of instrument to be mounted onto the tripod
- The environment in which the tripod is to be erected
- The location of the reference point over which the tripod is to be mounted

Step 2 Find a tripod with a mounting plate suitable for mounting the chosen instrument.

Step 3 Find a plumb bob with enough string to use with the selected tripod.

Step 4 Verify the following on the selected tripod:
- That the tripod is undamaged and usable
- That the leg adjustments will both loosen and tighten properly

- That the adjustable legs can be adjusted and secured to hold the weight of the tripod and the instrument
- That the mounting plate appears to be usable for the instrument chosen

9.3.0 Leveling Rods

Two people are required when a conventional leveling instrument is used; the first operates the instrument, and the second holds a vertical measuring device, called a leveling rod (*Figure 18*), in the area where the grade or elevation is being checked. Leveling rods are made in many sizes, shapes, and colors. They can be made of wood, fiberglass, metal, or a combination of these materials. Leveling rods consist of two or more movable sections, allowing the rod to be adjusted to different lengths. Telescoping rods are also available.

STANDARD ROD (SAN FRANCISCO STYLE)   TELESCOPING ROD

101F18.EPS

Figure 18 ◆ Leveling rods.

Many styles of leveling rods are given geographic names, such as Philadelphia rods, Chicago rods, San Francisco rods, and Florida rods. The Philadelphia rod is a two-section rod with scales on both the front and back, which can be extended to about 13'. The Chicago and San Francisco rods consist of three sliding sections, with the Chicago rod being 12' long and the San Francisco rod available in several lengths. The Florida rod is a 10' long rod graduated with alternating 0.10' wide red and white stripes.

9.3.1 Reading Leveling Rods

There are two types of leveling rods in common use in the United States: direct-reading architect's rods and engineer's rods. Metric rods are also available. An architect's rod is graduated in feet, inches, and eighths of an inch (*Figure 19*). As shown, each line and space on an architect's rod is ⅛" wide. An engineer's rod is marked in feet, tenths of a foot, and hundredths of a foot. As shown in *Figure 19*, each line and space marked on an engineer's rod is ¹⁄₁₀₀' wide.

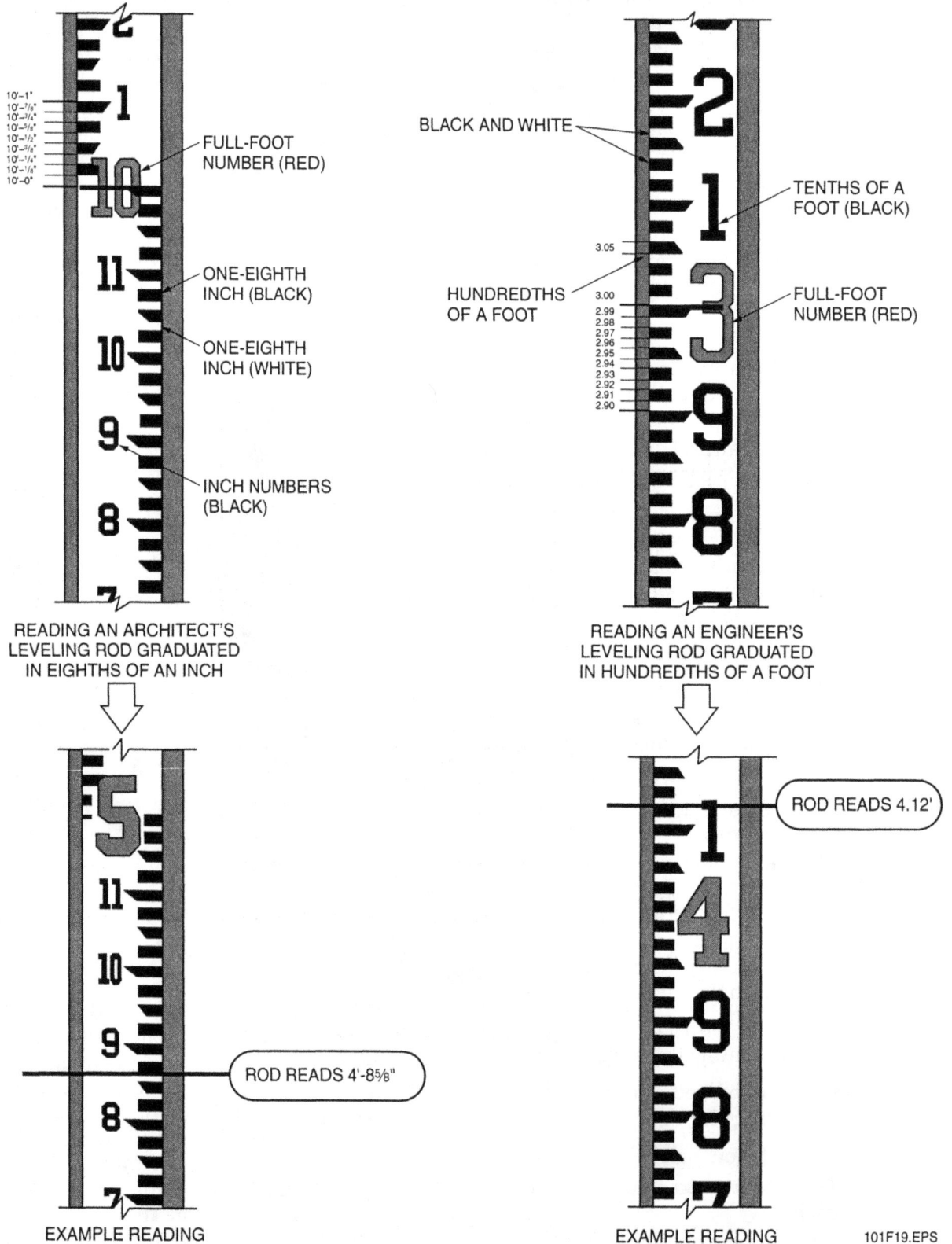

Figure 19 ◈ Reading a leveling rod.

9.3.2 Direct Elevation Rods

The direct elevation rod (*Figure 20*), commonly called a Lenker rod, is used when making grade and elevation measurements.

The direct elevation rod is made in two sections, allowing extension up to 10'. The rod has a continuously graduated metal ribbon or band that forms a continuous loop marked in either feet and inches for an architectural Lenker rod or feet and hundredths of a foot for the engineer's rod. The back section has a clamp to hold the rod in extension and a latch for locking the band in any

required position. The numbers on the band are read downward from the top of the rod. After properly setting the band position on a known bench mark, all subsequent rod readings will be the elevations of the points on the ground where the rod is held. The advantage of this is that no additions or subtractions of **backsight (BS)** or **foresight (FS)** readings are necessary.

An example procedure for using a direct elevation rod is shown in *Figure 21* and discussed below.

Step 1   Set up a leveling instrument so that its line of sight/laser beam can intersect a rod held on a known bench mark, plus two or more additional elevation points.

Step 2   Place the rod on the known bench mark. For our example, this is assumed to be 1,005.6' (*Figure 21*, position A).

Step 3   Loosen the band knob and move the rod band until the leveling instrument's line of sight or beam intersects the last two digits of the known bench mark elevation, then clamp the rod band securely in this position. For our example, this position is at 5.6'.

> **NOTE**
>
> As elevations increase or decrease, it may be necessary to extend or shorten the rod in order to bring the rod band in view. This is done by pressing the knob and moving the front section of the rod up or down as required.

Step 4   Move the rod to a new location (position B), then read the elevation directly on the band. For our example, this is 7.6' or 1,007.6' (by prefixing the 1,000). Thus, the elevation of this point is 2' higher than the bench mark (1,007.6' – 1,005.6' = 2').

Step 5   Move the rod to a new location (position C), then read the elevation directly on the band. For our example, this is 2.6' or 1,002.6' (by prefixing the 1,000). Thus, the elevation of this point is 3' lower than the bench mark (1,005.6' – 1,002.6' = 3').

Step 6   Repeat Step 5 as required to measure other elevation points

9.3.3 Leveling Rod Accessories

Several accessories are used with leveling rods. A movable red and white metal disk called a target (*Figure 22*) is used to help make more precise rod readings. The target's vernier scale is set parallel

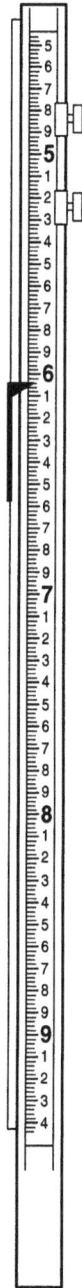

101F20.EPS

Figure 20 ◆ Direct elevation rod.

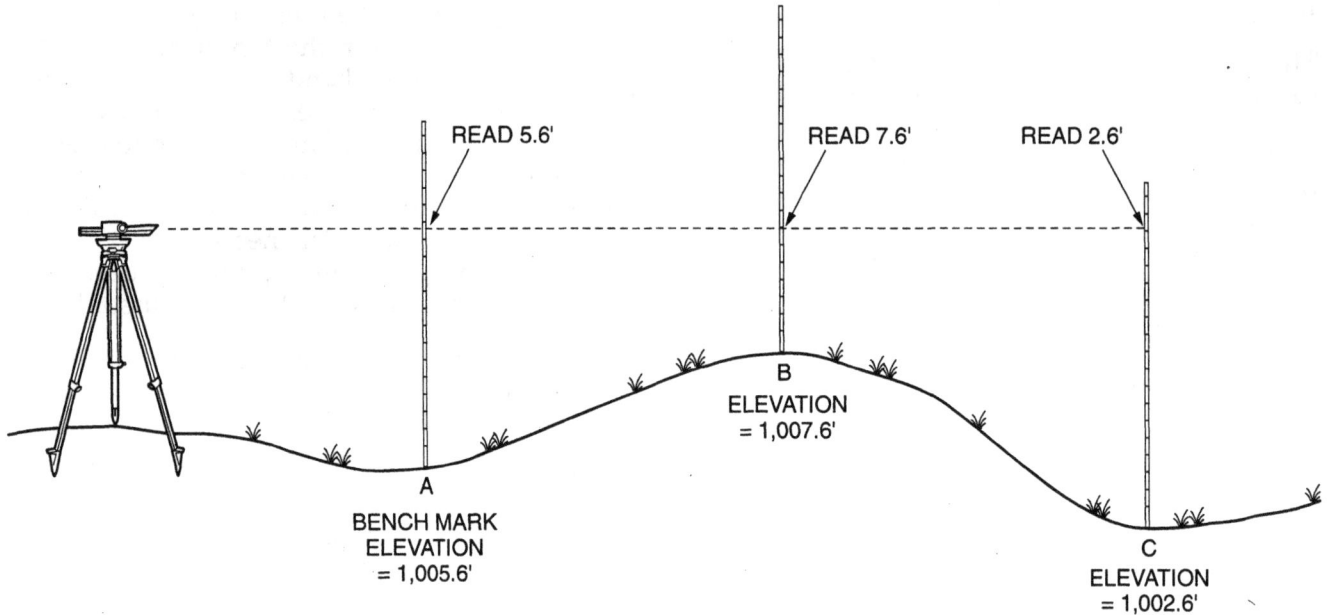

Figure 21 ◆ Using a direct elevation rod.

to and beside the primary scale of the leveling rod. Its use enables readings to the nearest sixty-fourth of a foot (architect's rod) or nearest thousandth of a foot (engineer's rod). The target is moved up or down on the rod until the 0 on the vernier scale is lined up with the crosshairs of the leveling instrument. The target is then clamped in place. To read the vernier scale, count the number of vernier divisions up from the 0 (index mark) until one of the vernier divisions lines up exactly with a division on the rod scale itself. This number is added to the last division on the rod, just below the vernier's index mark.

A bull's-eye rod level is normally attached to a leveling rod for use in keeping the rod plumb for sighting and while the reading is being taken.

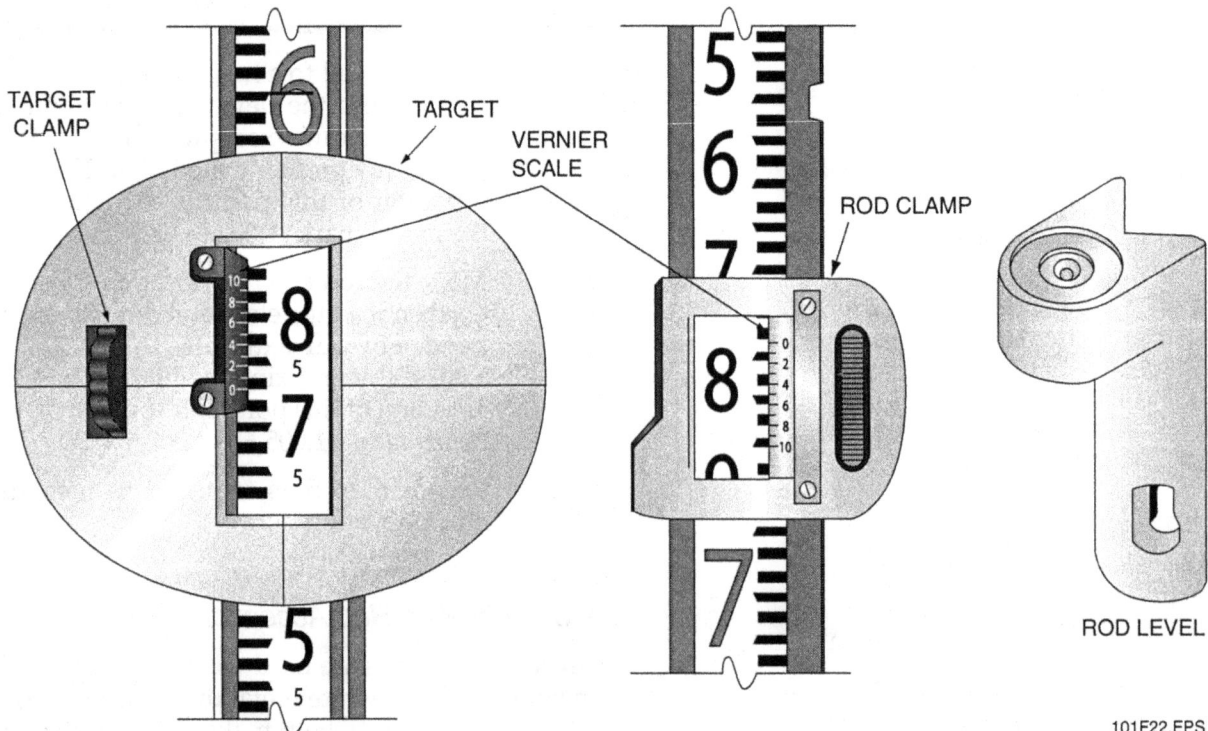

Figure 22 ◆ Rod accessories.

9.3.4 Care and Handling of Leveling Rods

Leveling rods are made to withstand the severity of everyday use, but they must be handled, stored, and used properly to avoid unnecessary damage. A damaged rod can contribute to errors.

Guidelines for the proper care and handling of leveling rods are as follows:

- Clean the face, joints, and bottom of the rod frequently during use.
- Avoid touching the face of the rod. Over time, this can cause the numbers and markings to be worn off.
- Make sure all the rod hardware is securely fastened.
- When using a telescoping-type leveling rod, make sure that it is fully extended. Failure to extend a rod fully will result in major errors.
- Never throw a leveling rod into the back of a truck or leave it sticking out from a truck. When not in use, store it in its protective case.

10.0.0 ◆ PLANS

Planning the layout of structures and/or other features for a construction project is an important task. In addition to designing and planning the project structures or features, planning must also take into consideration prevailing codes, **covenants**, planning and zone regulations, site utility **easements**, and other requirements. Project planning is normally done by architectural and engineering firms. Once the initial planning is done, the details for the job, including those pertaining to site layout, are documented in the form of a drawing set and specifications. *Figure 23* shows the drawing content of a typical drawing set used for a construction project. Details about specifications and each different kind of drawing contained in a drawing set are covered later in your training in the *Blueprint Reading for Surveyors* module.

11.0.0 ◆ SAFETY

Performing layout work at a construction site can be hazardous if proper safety precautions and rules are not followed. Numerous hazards can exist on a construction site. They can occur because of natural causes, inclement weather, insects or animals, highways, and the construction itself. Because each site layout task is typically performed by no more than two or three persons, each person must be capable of working unsupervised, including being responsible for his or her own job-site safety. General safety guidelines to be followed when working on a construction site were covered earlier in the Core Curriculum module *Basic Safety*. Some safety

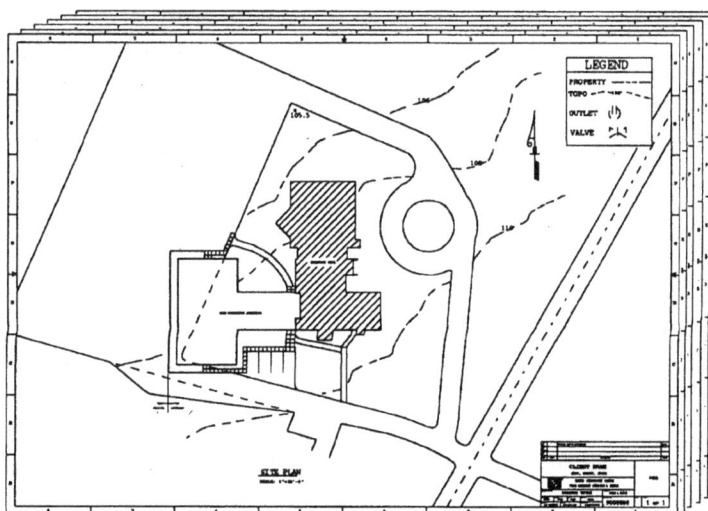

TITLE SHEET(S)
ARCHITECTURAL DRAWINGS
- SITE (PLOT) PLAN
- FOUNDATION PLAN
- FLOOR PLANS
- INTERIOR/EXTERIOR ELEVATIONS
- SECTIONS
- DETAILS
- SCHEDULES

STRUCTURAL DRAWINGS
PLUMBING PLANS
MECHANICAL PLANS
ELECTRICAL PLANS

101F23.TIF

Figure 23 ◆ Typical drawing set.

guidelines that apply especially to site layout and surveying work are given here:

- Dress properly. Wear clothing appropriate for the season. The clothing being worn should not be too tight or too loose. Make sure to wear bright-colored safety vests when working in areas subjected to vehicle or construction equipment traffic. Wear high boots and other appropriate clothing when working in brush or other uncleared areas that may be inhabited by snakes, insects, or other pests. Wear gloves when working in brush that may contain briars or poison ivy.

- Know the symptoms and how to avoid seasonal exposure hazards, such sunstroke and heat exhaustion in the summer or hypothermia and frostbite in the winter.

- Make yourself aware of what is going on at the job site. Prior to each day's work, find out when and where any exceptionally hazardous operations, such as blasting, are scheduled to take place. If you must work in an area while construction vehicles and equipment are also operating, notify the operator of the vehicles and equipment of your presence in the area.

- Do not enter ditches or other excavations that have not been properly shored or otherwise protected from cave-ins.

- Avoid working in open areas when thunderstorms are approaching or during thunderstorms.

- Do not climb over fences and similar obstacles while carrying equipment.

- Do not look at the sun through instrument telescopes unless special filters are used; otherwise, serious and permanent eye damage can occur.

- Be careful not to allow leveling rods to come into contact with overhead electrical wires or steel tapes to be dragged across extension cords.

- Make yourself aware of the locations of all open shafts or open sides of a multi-story building within your immediate work area. It is easy to fall into or out of these openings during such activities as making measurements or backing up. If no barriers are placed around openings, they should be put in place before proceeding with your work.

- Do not enter confined areas until your entrance has been approved by the proper authorities. Ditches, tunnels, storm sewers, or other confined or closed areas may contain hazardous vapors.

- Make sure all members of your site layout crew know the radio frequencies or channels and/or hand signals being used for communication among crew members.

12.0.0 ◆ CAREERS IN SITE LAYOUT

Employment opportunities in the field of site layout are affected by trends in construction. As the economy continues to grow, more site layout technicians will be needed to meet the demand for new construction. Today, the demand for qualified site layout technicians is growing and is expected to outpace the supply of skilled workers.

A career in site layout provides an opportunity to work both outdoors and indoors. It also offers the opportunity to travel and work in all parts of the United States and in foreign countries, in both rural and metropolitan areas. Technicians who perform site layout work are involved in heavy construction that typically requires a great deal of earthwork, substantial concrete placement, and the installation of mechanical and structural steel.

Some companies refer to people who do site layout work as field engineers. This job title does not imply that the individual is a graduate of an engineering school or has any type of professional registration. Its intent is to describe someone who performs technically oriented site layout tasks.

As with any career, a career in site layout has many steps and many options. A career in site layout usually begins by entering a site layout apprenticeship program, such as this one, that prepares you to become a site layout journeyman. The specific progression from entry-level apprentice to journeyman can vary depending on the particular apprenticeship program and your employer's specific needs. A typical progression involving increasing levels of experience and achievement is outlined here:

- *Level one* – After completing level one, the trainee will possess a basic knowledge of site layout field tasks and operations and the different types of surveys. The trainee is able to do the following:

  – Name the different types of surveys and explain the basic differences between them.
  – Name the types of survey control points and explain their differences.
  – Perform basic math and measurements related to surveying and site layout tasks.
  – State the names, purpose, parts, setup, transport, and calibration requirements for selected types of surveying instruments.
  – Correctly handle, set up, clean, and make basic measurements using selected surveying tools and equipment.
  – Make entries in surveying field notes.
  – Under supervision, perform common tasks such as line clearing, establishing points, taping, leveling, and compass reading.
  – Read and interpret basic construction drawings and specifications.

- *Level two* – The trainee builds on the knowledge and skills learned in level one. After completing level two, the trainee is able to do the following:
  - Demonstrate the principles of performing basic surveys.
  - Operate, check, and make field adjustments to transits, theodolites, total stations, data collectors, and similar instruments.
  - Perform more complex surveying computations.
  - Demonstrate a basic familiarity and understanding of data collectors, handheld calculators, PDAs, and computer operating systems.
  - Enter field data and produce positional information using a handheld calculator or computer software.
  - Interpret control point records and data sheets and locate points in the field.
  - Record surveying data in a neat and organized manner in field notes.
  - Under supervision, coordinate field work for a variety of standard surveys.

For those who want to advance beyond the site layout journeyman level, opportunities exist to become a construction site surveyor, construction superintendent, or project manager. Many site layout technicians decide to further their education and training in order to become registered surveyors. In addition to working for construction-related companies, many site layout persons work for the following types of organizations:

- Surveying and engineering firms
- Departments of transportation
- Departments of public works
- Energy and utility companies
- Bureaus of land management
- U.S. forest service
- Departments of agriculture

13.0.0 ◆ TRAINING

Most employers consider two basic forms of training programs. The primary one is on-the-job training (OJT) which improves competence of their employees in order to provide better customer service and for the continuity and growth of the company. The second is formal apprenticeship training, which provides the same type of training, but also conforms to federal and state requirements under the *Code of Federal Regulations (CFR)*, *Titles 29:29* and *29:30*.

13.1.0 Standardized Training by NCCER

The National Center for Construction Education and Research (NCCER) is an independent, private educational foundation founded and funded by the construction industry to solve the training problems plaguing the industry today. The basic idea of NCCER is to replace governmental control and credentialing of the construction workforce with industry-driven training and education programs. NCCER departs from traditional classroom or distance learning by offering a competency-based training regimen. Competency-based training means that instead of simply requiring specific hours of classroom training and set hours of OJT, you have to prove that you know what is required and successfully demonstrate specific skills.

All completion information on every trainee is sent to NCCER and kept within the National Registry. NCCER can confirm training and skills for workers as they move from company to company, state to state, or to different offices in the same company. These are portable credentials and are recognized nationally.

13.2.0 Apprenticeship Training

As stated earlier, formal apprenticeship programs conform to federal and state requirements under *CFR Titles 29:29* and *29:30*. All approved apprenticeship programs provide OJT as well as classroom instruction. The related training requirement is fulfilled by NCCER craft training programs. The main difference between NCCER training and registered apprenticeship programs is that apprenticeship has specific time limits in which the training must be completed. Apprenticeship standards set guidelines for recruiting and outreach and a specific time limit for each of a variety of OJT tasks. Additionally, reporting requirements and audits ensure adherence to the apprenticeship standards. Companies and employer associations register their individual apprenticeship programs with the Apprenticeship, Training, Employer, and Labor Services (ATELS) within the U.S. Department of Labor and, in some instances, with state apprenticeship councils (SAC). OJT of 2,000 hours per year and a minimum of 144 hours of classroom-related training are required. Apprenticeship programs vary in length from 2,000 hours to 10,000 hours.

13.3.0 Youth Training and Apprenticeship Programs

Youth apprenticeship programs that allow students to begin their apprenticeship or craft training while still in high school are available. A student entering the program in the 11th grade may complete as much as one year of the NCCER training program by high school graduation. In addition, programs run in cooperation with local construction industry employers allow students to work in the craft and earn money while still in school. Upon graduation, students can enter the industry at a higher level and with more pay than someone just starting in a training program. Students participating in NCCER or youth apprenticeship training are recognized through official transcripts and can enter the second level or year of the program wherever it is offered. They may also have the option of applying credits at two-year or four-year colleges that offer degree or certificate programs in their selected field of study.

14.0.0 ◆ SITE LAYOUT PERSONNEL CREED AND RESPONSIBILITIES

As a site layout technician, you play a critical role in the successful completion of each project assigned to you. You wear many hats and interact with everyone on the job site. Typically, pressure is applied from every direction to do fast and accurate layouts. This pressure acts as a reminder of the importance of your daily tasks. As you advance in the trade, you will be required to make decisions on a daily basis that have a direct impact on a project's budget and schedule. Making engineering mistakes in today's fast and competitive market can easily mean the difference between profit and loss on a job, especially one with a tight budget.

Sometimes it is hard to remain confident in your abilities when others think that the job you work so hard to do professionally is easy or not important. You must have patience with others if they inadvertently or carelessly destroy your work, such as can happen with the layout for a line or bench mark. You need to understand that people in other trades often do not realize how much time and effort it takes to accurately locate a line or bench mark.

The supervisor of a site layout crew is the person that everyone goes to for support and answers. This person and others in the crew must keep an open mind and remain calm in response to different and difficult situations. Staying courteous and professional when other people around

you are losing their tempers helps keep the situation under control. When you become angry and unsure of what you are doing, it tends to panic everyone. Sometime it is hard not to get angry when others insist that your work is wrong, requiring you to prove that it is right. Some people will try to find fault with your work in order to hide their own mistakes. When you lose your temper and your composure, you lose your ability to reason and solve problems, and that's your job.

The supervisor of a site layout crew is usually the project management's contact for gathering and relaying accurate information in the field. On a daily basis, the site layout crew supervisor communicates with management, subcontractors, inspectors, and other craftspeople.

As a site layout technician, you travel all over the job site and come in contact with many people. Be professional in your appearance and mannerisms at all times because first impressions are important. A person who shows little respect for himself or herself is not going to be taken seriously or trusted by others to do a professional job. A sloppy appearance, bad attitude, use of offensive language, and/or poor communication skills all reflect on both you and your employer.

As a site layout technician, you should be a role model for safety. Your work is performed around moving heavy construction vehicles and equipment, in foundations, in trenches, on the edges of buildings, around floor openings, and up on columns and walls. For this reason, when doing a job, your focus must remain equally on safety as well as on the job at hand.

Procrastination and being poorly prepared contribute to making mistakes and working inefficiently. Knowing when and what you need to do for each task or job is important. For each task or job, have a plan of action, and have everyone on the crew understand their role in the plan. Be sure to identify and have available the equipment and tools needed to do the job. A crew standing around waiting for instructions and/or equipment wastes time and money. Sometimes it may be necessary to work through lunch or stay late in order to have the time to plan or check your work properly.

Working with honesty, quality, and efficiency earns the trust and respect of everyone and sets a good example for others to follow. Trust and respect are something that, once lost, are almost impossible to regain.

Your success depends on taking personal responsibility for making sure that your work is done correctly. A good site layout technician never just gives line or grade and then hopes that the work is going to turn out right. To avoid mistakes, you have to continually check your work, and often the work of others as well. For example, if you have laid out the lines for a concrete slab correctly, but the concrete ends up in the wrong place, you haven't done your job.

To be a successful site layout technician, you need to constantly research and better yourself. You have to be able to identify problems and solve them. The hours are sometimes long and the working conditions difficult. However, the experience gained gives you an understanding and broad knowledge of how all of the physical components and various skills involved in a construction project relate to each other. It also gives you a sense of pride and self worth and a strong technical foundation upon which to build a career. The knowledge, self-confidence, and dedication to teamwork gained not only helps you on the job, but in all that you do.

15.0.0 ◆ COMMUNICATING AS A PROFESSIONAL

It is rare to have a job in which people skills do not count. The way you interact at the job site can make or break your success in this field. Most skilled site layout technicians recognize that good people skills are valuable tools of the trade. After all, first impressions count. It is not enough just to be technically competent. You must show those in your trade, other trades, and the customer that you will handle the job and solve any problems efficiently and pleasantly. Every interaction is an opportunity to convey a professional image for you and your company.

Professionals pay attention to all the products of their work, including communication. They always act in a professional and ethical manner. This means demonstrating the traits of honesty, productivity, safety, cooperation, and civility. Just as it is dangerous to ignore safety concerns, it is also risky to ignore how your words and actions may be interpreted by others.

You will be working with many different people. In one day, you could interact with a supervisor, co-workers, other tradespeople, customers, and an inspector. Every work site is different. You may be the lone technician or part of a larger team. When communications are positive and effective, the stage is set for a productive working environment. How you treat other people is your personal responsibility. The way you get treated is often a result of how well you interact with others.

Experienced workers realize that success in the workplace is determined by a combination of skill, knowledge, and behavior appropriate to the situation. It is important to convey a professional image. The level of professionalism demonstrated by each skilled worker reflects back on the industry as a whole. Clear, effective communication makes work more productive and pleasant for everyone. If your natural people skills are not what they should be for this kind of work, take the time to learn how to improve them.

15.1.0 Interpreting Instructions

Interpreting and recalling instructions are very common tasks in every workplace. The misinterpretation of oral, written, or graphic instructions can lead to costly mistakes. Mistakes can negatively affect safety, quality of work, and the confidence others have in you to do the job correctly.

Communication studies show that the words or content of a face-to-face interaction influence only about 7% of how the total communication is perceived. Body language (55%) and tone of voice (38%) provide over 90% of the speaker's message. These are the clues we look for when interpreting a message (*Figure 24*). Usually, we pick up on many things besides the actual words spoken. There are even more opportunities for misinterpretation when people communicate by phone or in writing. Just think of how much facial expressions and hand gestures contribute to your understanding of what is being said.

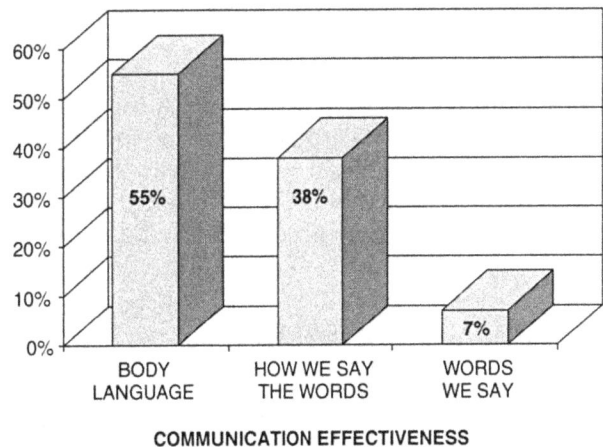

COMMUNICATION EFFECTIVENESS

101F24.EPS

Figure 24 ◆ How communication is interpreted.

Good communication skills include active listening and the precise presentation of information. They are both essential for efficient work practices and avoiding costly mistakes. These people skills are not easy to achieve without a good deal of effort and practice.

Everyone has stories about how they were treated as a customer by a representative of a particular company. The situations that were most

---

frustrating are usually the ones where you, as the customer, were not treated with respect. The situations that left you with a positive feeling were very different. If you recall those interactions, you may remember that the employee acted confident, helpful, and professional. Most likely, the service person appeared to really concentrate and listen carefully to what you were asking. When he or she gave you information, it was probably presented clearly and concisely. Even if the situation itself was frustrating, the way you were treated as a customer made a difference. If a problem was handled well, you are more likely to continue to do business there.

The first rule of good communication is to concentrate on the speaker and actively listen to what is being said. This means not interrupting and avoiding the tendency to think about what you are going to say next. There is plenty of time to think of what to say after you are sure you have understood the message. Because what is being said could be open to interpretation, it helps to restate what you think you have been told. Restating the important parts of the message will help you remember it better. When you hear the information and repeat it in your own words, it gets placed into your memory twice. Writing the information down is a third way to keep it in focus.

The key to preventing a misunderstanding is to check out the most important information. Restate what you think are the most important points and watch how the speaker responds. Do not worry about sounding stupid. It is better to get the correct message before you start a job. If there is any doubt about what the speaker meant to say, asking questions is the best way to clear things up. The answer could mean the difference between doing it right the first time and having to waste time on rework.

15.2.0 Restating Written Instructions Orally

Written instructions can often be very short. You may have to read between the lines to understand the full message. The notes you read may be incomplete or difficult to understand. It is always better to get the correct information up front. This is especially true when instructions are critical to a successful installation or follow-up. Again, the trick is to restate the written instructions orally in your own words to check for accuracy.

15.3.0 Clearly Relaying a Customer Requirement

Customers may ask you questions or request that you tell a co-worker or supervisor a requirement they have for the job. To convey the information accurately, be sure you understand what the requirement is first. Briefly restate the basics of the requirement back to the customer. He or she will be able to clear up any misunderstandings before you relay the message. If you cannot answer a customer's question or solve a problem, direct the customer to your supervisor or someone who can.

When relaying a message, be sure you have the co-worker or supervisor's full attention. Telling a person something important when they are distracted guarantees that only part of the message will be heard. Even less will be remembered. Gain their full attention by stating why the requirement is important and how it involves them. State the requirement clearly, using a logical sequence. Describe the desired outcome first, then detail the steps required.

State the most important information up front, then repeat it. If the message is complicated, be sure the listener understands each part before going on to the next. You can check for understanding by asking the listener to tell you what they understand about the requirement. Use this as an opportunity to correct any misinformation before mistakes are made.

Finally, follow up to make sure the customer's needs are fulfilled or questions have been answered. Later, ask your co-worker what action was taken. Find out if the customer knows the requirement was acted upon or if delays have been explained. The extra effort you make to get back to the customer will increase their confidence in you as a professional. The customer usually does not care who solves the problem, just that it was solved.

Summary

A career in site layout provides the opportunity to work both outdoors and indoors and travel worldwide. This usually is done on jobs involving heavy construction and require a great deal of earthwork, substantial concrete placement, and the installation of mechanical and structural steel. Site layout technicians typically perform the following tasks:

- Survey and gather information about a proposed job site
- Lay out the location of structures and other features on the site
- Check the dimensions of the structures as they are being built
- Document the completed work
- Verify that all the work is done in accordance with the design plans and specifications

Many job and career opportunities are related to the field of site layout. An apprenticeship program that combines competency-based, hands-on training with classroom instruction has proven to be the most effective means for a person to learn and advance in a specialized trade such as site layout. Developing job skills is only part of the solution; it is just as important to learn good work habits, convey a positive, cooperative attitude to those around you, and practice good safety habits every day.

Review Questions

1. Earth's curvature must be taken into consideration and the measured distances and angles adjusted accordingly when performing a _____ survey.
   a. plane
   b. construction
   c. geodetic
   d. topographical

2. Locations in the United States mainland have a _____ latitude and a(n) _____ longitude.
   a. north; west
   b. south; east
   c. north; east
   d. south; west

3. Site layout technicians should never make any measurements that locate _____.
   a. footings and foundations
   b. working control points
   c. layout of embedded items
   d. property lines or boundaries

4. The second network of control points that are established when laying out control points at a construction site are called _____.
   a. secondary control points
   b. primary control points
   c. reference azimuth control points
   d. building layout/working control points

5. All of the following are recommended practices for marking information on stakes *except* _____.
   a. use all capital letters
   b. do not crowd words and numbers
   c. use abbreviations whenever possible
   d. mark the main information on the direction of use

6. In surveying and site layout work, mistakes _____.
   a. do not happen
   b. can be eliminated by checking and rechecking your work
   c. follow the laws of probability
   d. occur as a result of limitations in surveying equipment

7. Pulling too hard on a steel tape when taping will eventually stretch and permanently damage the tape.
   a. True
   b. False

8. If four 100' tape measurements were previously made when taping a distance, and the last measurement recorded is 50.25', the total distance measured is _____.

9. A 100' tape is being used to lay out a 300' wall of a building. The temperature of the tape is 80°F. What distance, as indicated by the tape, needs to be laid out to set the point at the required distance?
   a. 299.93'
   b. 299.98'
   c. 300.00'
   d. 300.02'

10. A distance between two points is 250'. If measuring this distance with a 100' tape on a 50°F day, what distance would you expect to read directly from the tape?
    a. 249.07'
    b. 250.00'
    c. 250.03'
    d. 250.08'

Refer to *Figure 1* when answering Questions 11 and 12.

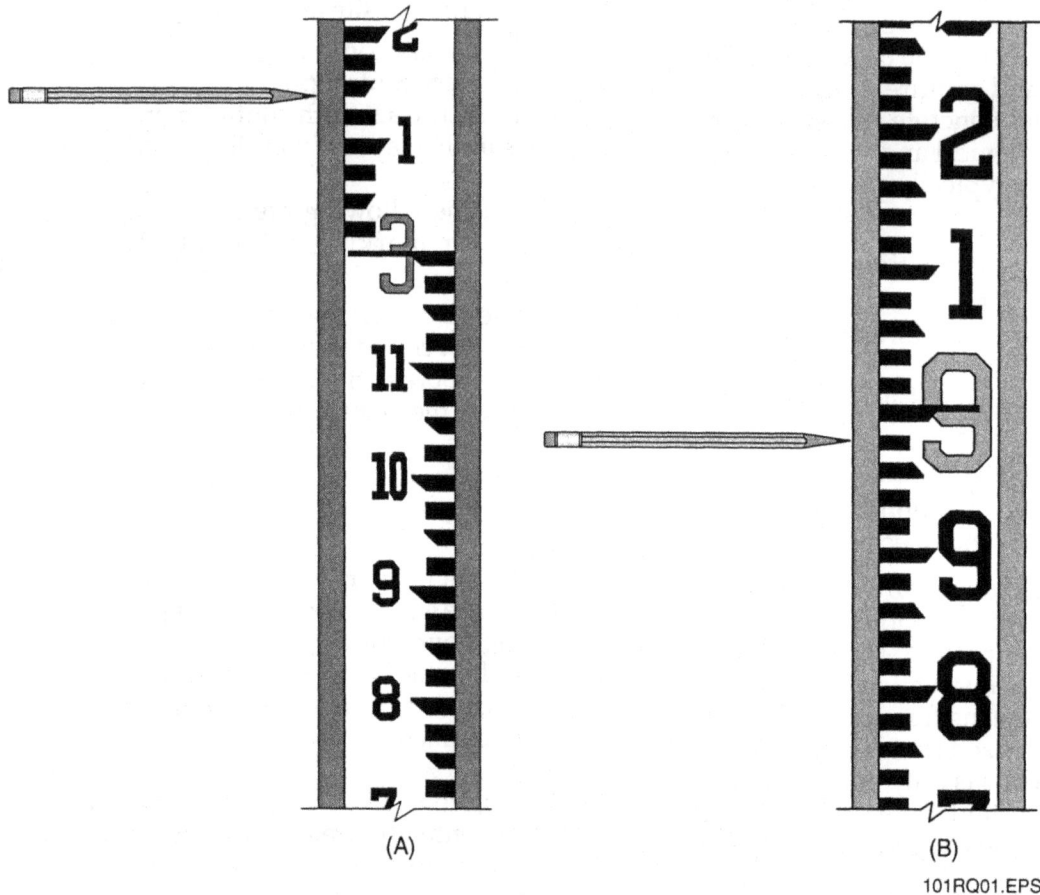

Figure 1

101RQ01.EPS

11. As indicated at the pencil point, the architect's leveling rod shown in *Figure 1A* reads _____.

    a. 3.14'
    b. 3'-1⅛"
    c. 3'-1⅜"
    d. 3.05'

12. As indicated at the pencil point, the engineer's leveling rod shown in *Figure 1B* reads _____.

    a. 8'-9⅞"
    b. 9'-⅞"
    c. 8.98'
    d. 8.99'

13. When using a direct elevation (Lenker) rod to measure the elevation of an unknown point, the rod reads 6.5'. If the rod was initially adjusted to a bench mark elevation of 1002.5', the elevation for the unknown point is _____.

    a. 1004.5'
    b. 1004.0'
    c. 1006.5'
    d. 1009.0'

14. A competency-based training program is one that requires the trainee to _____.

    a. receive at least four years of classroom training
    b. receive on-the-job training for at least two years
    c. pass a series of written tests
    d. demonstrate the ability to perform specific job-related tasks

15. When having a face-to-face conversation with others, the words you use account for _____ of how the total communication is perceived.

    a. 7%
    b. 50%
    c. 75%
    d. 90%

Trade Terms Introduced in This Module

Backsight (BS): When measuring angles, the sight taken with the instrument telescope on a point or line being used as the starting point for an angle measurement. Also, a reading taken on a leveling rod held on a point of known elevation to determine the height of the leveling instrument.

Bench mark: A permanent reference point with a precisely determined elevation.

Breaking the tape: Making measurements using a portion of a full tape's length in a series of steps.

Control points: A network of correlated horizontal and vertical points that serve as a common set of reference points at a job site.

Contour lines: Lines on a map that join points of the same elevation.

Covenants: Binding agreements made by two or more individuals or parties to do or not do a specific task or thing.

Crosshairs: A set of lines, typically horizontal and vertical, placed in a telescope used for sighting purposes.

Datum: A permanent horizontal and vertical reference for a project.

Easement: A legal right-of-way provision given to one person or entity on another person's or entity's property.

Foresight (FS): When measuring angles, the sight taken with the instrument telescope on the ending point or line for an angle measurement. Also, a reading taken on a leveling rod held on a point in order to determine a new elevation.

Geodetic surveys: Surveys in which Earth's curvature is taken into consideration.

Grade: The vertical (elevated) position of an object or feature above a known datum.

Line: Two connected points referenced to a project datum.

Mean sea level (MSL): A reference datum commonly used when measuring heights and depths. MSL is assigned the elevation of 0.000 feet (or meters), allowing all points on Earth to be described by elevations above and below zero.

Occupational Safety and Health Administration (OSHA): An agency of the U.S. Department of Labor that is responsible for developing and enforcing regulations to support the *Occupational Safety and Health Act,* the federal law designed to protect people from unsafe work environments.

Plane surveys: Surveys in which Earth's curvature is not taken into consideration.

Surveying: The process of determining the location of physical land and/or man-made features relative to one another and relative to a predetermined reference on the surface of Earth.

Random error: An inadvertent error introduced into all measurements because of human limitations. Random errors follow the laws of probability; therefore, multiple random errors tend to cancel each other out.

Relief: In surveying, the differences in the height of land forms in any particular area.

Setback: The distance between a building or other structure and the building site property line. Local zoning regulations or construction codes normally specify the minimum setbacks permitted for buildings.

Systematic error: An error that occurs in the same magnitude and the same sign for each measurement of a distance, angle, or elevation.

Traverse: The measurement of a continuous series of angles or distances.

Additional Resources

This module is intended to be a thorough resource for task training. The following reference works are suggested for further study. These are optional materials for continued education rather than for task training.

Construction Surveying and Layout, Westley G. Crawford, West Lafayette, IN: Creative Construction Publishing, Inc.

Principles and Practices of Commercial Construction, Cameron K. Andres and Ronald C. Smith, Upper Saddle River, NJ: Prentice Hall.

Figure Credits

Cianbro Corporation	101F02
TOOLZ, Inc.	101F15
Sokkia Corporation	101F18

CONTREN® LEARNING SERIES — USER UPDATES

The NCCER makes every effort to keep these textbooks up-to-date and free of technical errors. We appreciate your help in this process. If you have an idea for improving this textbook, or if you find an error, a typographical mistake, or an inaccuracy in NCCER's Contren® textbooks, please write us, using this form or a photocopy. Be sure to include the exact module number, page number, a detailed description, and the correction, if applicable. Your input will be brought to the attention of the Technical Review Committee. Thank you for your assistance.

Instructors – If you found that additional materials were necessary in order to teach this module effectively, please let us know so that we may include them in the Equipment/Materials list in the Instructor's Guide.

Write: Product Development
National Center for Construction Education and Research
P.O. Box 141104, Gainesville, FL 32614-1104

Fax: 352-334-0932

E-mail: curriculum@nccer.org

Craft _____ Module Name _____

Copyright Date _____ Module Number _____ Page Number(s) _____

Description _____

_____

_____

_____

(Optional) Correction _____

_____

_____

(Optional) Your Name and Address _____

_____

_____

Surveying Math

COURSE MAP

This course map shows all of the modules in the first level of the Site Layout curriculum. The suggested training order begins at the bottom and proceeds up. Skill levels increase as you advance on the course map. The local Training Program Sponsor may adjust the training order.

SITE LAYOUT LEVEL ONE

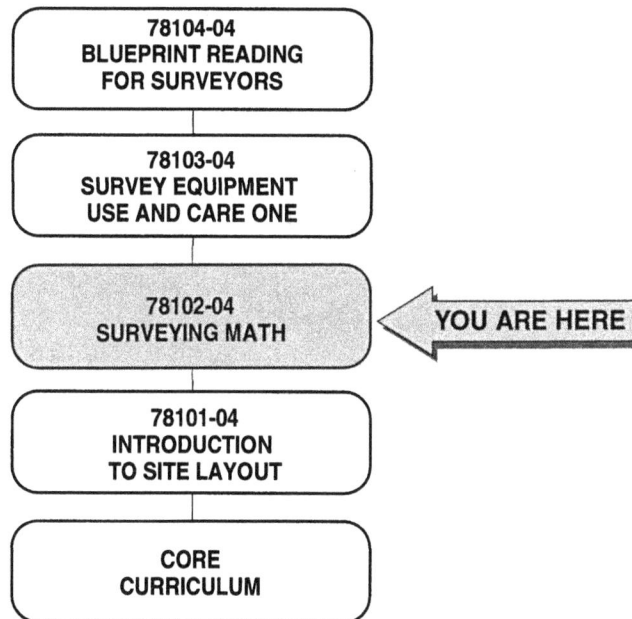

```
┌─────────────────────────┐
│       78104-04          │
│   BLUEPRINT READING     │
│    FOR SURVEYORS        │
└─────────────────────────┘

┌─────────────────────────┐
│       78103-04          │
│   SURVEY EQUIPMENT      │
│   USE AND CARE ONE      │
└─────────────────────────┘

┌─────────────────────────┐
│       78102-04          │        ◀═══ YOU ARE HERE
│   SURVEYING MATH        │
└─────────────────────────┘

┌─────────────────────────┐
│       78101-04          │
│    INTRODUCTION         │
│   TO SITE LAYOUT        │
└─────────────────────────┘

┌─────────────────────────┐
│        CORE             │
│     CURRICULUM          │
└─────────────────────────┘
```

102CMAP.EPS

Figures

Tables

Surveying Math

Objectives

When you have completed this module, you will be able to do the following:

1. Solve basic equations, including those involving squares and square roots.
2. Identify basic geometric shapes and angles.
3. Apply the Pythagorean theorem to solve math problems involving right triangles.
4. Perform decimal and metric conversions for linear measures, areas, and volumes.

Prerequisites

Before you begin this module, it is recommended that you successfully complete the following: Core Curriculum; Site Layout Level One, Module 78101-04.

Required Trainee Materials

1. Pencil and paper
2. Protractor
3. Calculator

1.0.0 ◆ INTRODUCTION

This module expands on the materials learned in *Introduction to Construction Math* in *Core Curriculum*. In that module, you studied whole numbers, fractions, decimals, and the metric system. If necessary, you may want to review all or part of the material covered in *Introduction to Construction Math* before proceeding with the material covered here.

This module introduces basic **formulas** and equations. Site layout work requires the use of mathematical formulas when working with linear and angular measurements. Basic decimal and metric conversions useful in performing surveys and in interpreting construction drawings are introduced. Also covered is working with geometric shapes and angles, solving math problems involving **right triangles** using the **Pythagorean theorem**, and basic volume calculations.

2.0.0 ◆ WORKING WITH FORMULAS AND EQUATIONS

You will solve many surveying and other problems by using formulas. Formulas are algebraic equations made up of letters that represent values and symbols (mathematical signs such as + and × that tell you what to do). Study the following formula for calculating the area of a circle:

$$a = \pi r^2$$

To read this formula, you would say, "Area is equal to pi times radius **squared**." When you see an expression such as πr^2 in a formula, it means that you must multiply the values. In this case, you would multiply the radius squared by π.

Some values in formulas are variables. Any number can replace a variable. In this sample formula, r is a variable. No matter what size a circle is, and thus no matter what its radius is, we can determine its area with this formula. Some values, however, are constant. Pi is a constant. Its value is always 3.14159, often rounded off to 3.14. Whenever you see the word pi or its symbol (π) in a formula, you know to replace it with 3.14.

Equations are collections of numbers, symbols, and mathematical operators connected by equal signs (=). Consider the following equation for calculating the area of a rectangle:

$$\text{Area} = L \times W$$

A coefficient is a multiplier. In this equation, L is the coefficient of W. It can also be written as LW, without the multiplication sign. No multiplication sign is required when the intended relationship between symbols and letters is clear. For example, 2L means two times L (2 is the coefficient of L).

2.1.0 Sequence of Operations

Complicated equations must be solved by performing the indicated operations in a prescribed sequence: multiply, divide, add, and subtract (MDAS). For example, the following equation can result in a number of answers if the MDAS sequence is not followed:

$$3 + 3 \times 2 - 6 \div 3 = ?$$

To come up with the correct result, this equation must be solved in the following order:

Step 1	Multiply:	$3 + \underline{3 \times 2} - 6 \div 3$
Step 2	Divide:	$3 + 6 - \underline{6 \div 3}$
Step 3	Add:	$\underline{3 + 6} - 2$
Step 4	Subtract:	$\underline{9 - 2}$
	Result:	7

2.2.0 Squares and Square Roots

The formula for the area of a circle is just one in which you will need to work with squared numbers. You will also work with formulas in which you must find the **square root** of a given number.

A square is the product of a number or quantity multiplied by itself. For example, the square of 6 means 6×6. To denote a number as squared, simply place the exponent 2 above and to the right of the base number. An exponent is a small figure or symbol placed above and to the right of another figure or symbol to show how many times the latter is to be multiplied by itself. For example:

$$6^2 = 6 \times 6 = 36$$

The square root of a number is the divisor which, when multiplied by itself (squared), gives the number as a product. Extracting the square root refers to a process of finding the equal factors which, when multiplied together, return the original number. The process is identified by the radical symbol $[\sqrt{\ }]$. This symbol is a shorthand way of stating that the equal factors of the number under the radical sign are to be determined. Finding the square roots is necessary in many calculations, including those involving right **triangles**.

For example, $\sqrt{16}$ is read as the square root of 16. The number consists of the two equal factors 4 and 4. Thus, when 4 is squared, it is equal to 16. The term squaring a number simply means multiplying the number by itself.

The number 16 is a perfect square. Numbers that are perfect squares have whole numbers as the square roots. For example, the square roots of perfect squares 4, 25, 36, 121, and 324 are the whole numbers 2, 5, 6, 11, and 18, respectively.

Squares and square roots can be calculated by hand, but the process is very time consuming and subject to error. Most people find squares and square roots of numbers using a calculator. To find the square of a number, the calculator's square key $[x^2]$ is used. When pressed, it takes the number shown in the display and multiplies it by itself. For example, to square the number 4.235, you would enter 4.235, press the $[x^2]$ key, then read 17.935225 on the display.

Similarly, to find the square root of a number, the calculator's square root key $[\sqrt{\ }]$ or $[\sqrt{\ }_x]$ is used. When pressed, it calculates the square root of the number shown in the display. For example, to find the square root of the number 17.935225, enter 17.935225, press the $[\sqrt{\ }]$ or $[\sqrt{\ }_x]$ key, then read 4.235 on the display.

> **NOTE**
> On some calculators, the $[\sqrt{\ }]$ or $[\sqrt{\ }_x]$ key must be pressed before entering the number.

2.3.0 Using Formulas to Solve Problems

To solve surveying problems using formulas, enter the measurements taken into the equation that makes up the formula. For example, let's say you need to know the area of a round access cover located in one area of a job site. You measure across the cover and determine that its diameter is four feet. Its radius (r) is half that, or two feet. To determine the area of the round access cover, you can plug this number into the formula for the area of a circle (area = πr^2) in place of the letter *r*. Look at the formula with the numbers that represent the variable and the constant plugged into the equation:

$$a = \pi r^2$$

$$a = (3.14)(2^2)$$

Notice that 3.14 and 2^2 have been placed inside parentheses. When numbers are enclosed in parentheses, you must finish any calculations inside each set of parentheses before completing any other calculations. In this case, you will first find 2^2 by multiplying 2×2:

a = (3.14)(4)

a = 12.56 square feet

The parentheses in this equation are a type of grouping symbol. Grouping symbols are just like punctuation in writing. Imagine how hard it would be to read this module if there were no periods or commas to tell you where to stop or pause. Grouping symbols help you make sense of an equation, just like punctuation helps you make sense of a sentence. Grouping symbols tell you which numbers belong together and which functions to perform first. It's important to pay attention to how terms are grouped in a formula and to do the calculations in the right order. In more complex problems, additional types of grouping symbols, such as square brackets [ ] and braces { }, are used.

3.0.0 ◆ GEOMETRIC SHAPES

Every site at which you work can be divided into geometric shapes. Structures and land formations are representative of such shapes. You learned basic geometric shapes in the *Core Curriculum*. We will review some of that material and expand on it here because basic geometry is important to working with the shapes you will observe and measure as a surveyor.

3.1.0 Elements of Plane Geometry

Geometry is a study of figures. It consists of two main fields: plane geometry and solid geometry. Plane geometry is the study of plane figures such as squares, rectangles, triangles, circles, and **polygons**. Solid geometry is the study of figures that occupy space, such as cubes, spheres, and other three-dimensional objects. The focus of this module is on the elements of plane geometry.

3.1.1 Points and Lines

A geometric point represents a position only. It has no dimensions such as length, width, or height. A point is an origin or beginning, such as a starting place for a circle. Two points are necessary to define a line, and three points may define a flat surface, or plane, in space.

A line has one dimension: length (*Figure 1*). A straight line is defined as the shortest distance between two points. A line with no end points has an unlimited, or infinite, length. A line that begins at one point and ends at another has a fixed length

and is called a **line segment**. Note that the word *line* is commonly used to mean line segment. A line having the same direction throughout its length is called a straight line. A broken line is a series of connected straight lines extending in different directions. A line that continuously changes direction is called a curved line.

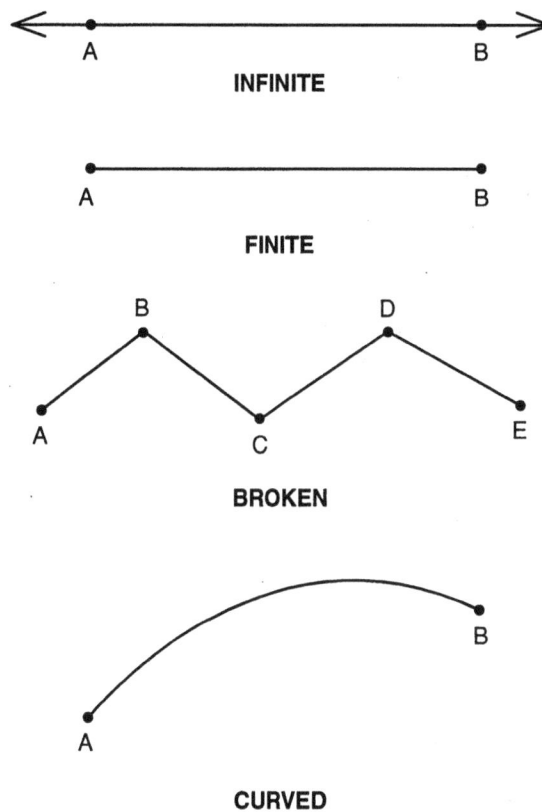

Figure 1 ◆ Lines.

A line that forms a right (90°) angle with one or more lines is said to be perpendicular to that/those lines (*Figure 2*). The distance from a point to a line is the measure of the perpendicular line drawn from that point to the line. Two or more straight lines that are the same distance apart at all perpendiculars are said to be parallel. Parallel lines do not intersect, even if extended infinitely.

3.1.2 Circles

A circle is a finite curved line that connects with itself (*Figure 3*). A circle also has these other properties:

- All points on a circle are the same distance, or equidistant, from the point at the center.
- The distance from the center to any point on the curved line, called the radius (r), is always the same.

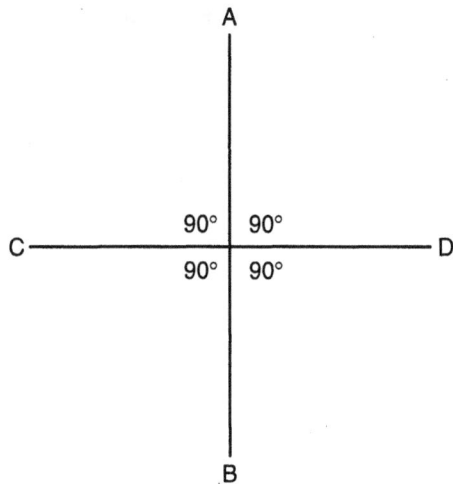

PERPENDICULAR LINES

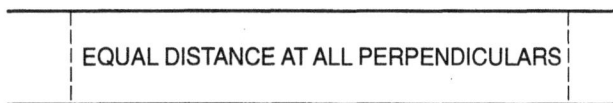

EQUAL DISTANCE AT ALL PERPENDICULARS

PARALLEL LINES

102F02.EPS

Figure 2 ◆ Perpendicular and parallel lines.

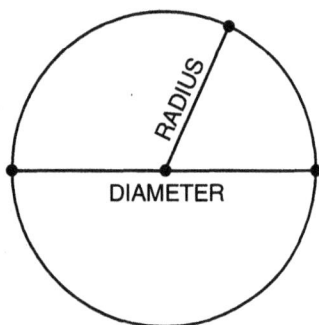

102F03.EPS

Figure 3 ◆ Circle.

- The shortest distance from any point on the curve through the center to a point directly opposite is called the diameter (d). The diameter is therefore equal to twice the radius (d = 2r).
- The distance around the outside of the circle is called the circumference. It can be determined by using the equation: circumference = πd, where π is a constant approximately equal to 3.14 and d is the diameter.
- A circle is divided into 360 parts with each part called a degree; therefore, one degree = $\frac{1}{360}$ of a circle. The degree is the unit of measurement commonly used in construction layout for measuring the size of angles.
- The total measure of all the angles formed by all consecutive radii equals 360°.

For surveying and site layout work, it is important to understand the following relationships relative to a circle: a full circle of angular rotation encompasses 360°. The degrees are further divided into minutes and seconds, where one degree is equal to 60 minutes (60') and one minute is equal to 60 seconds (60"). This arrangement allows angles to be expressed in terms of degrees, minutes, and seconds. This arrangement is called the sexagesimal system.

3.1.3 Angles

Two straight lines meeting at a point, called the vertex, form an angle (*Figure 4*). The two lines are the sides of the angle. These lines are called the rays of the angle. The angle is the amount of opening that exists between the rays. It is measured in degrees. Two ways are commonly used to identify angles. One is to assign a letter to the angle, such as angle D shown in *Figure 4*. This is written: ∠D. The other way is to name the two end points of the rays and put the vertex letter between them; for example, ∠ABC. When you show the angle measure in degrees, it should be written inside the angle, if possible. If the angle is too small to show the measurement, you may put it outside of the angle and draw an arrow to the inside.

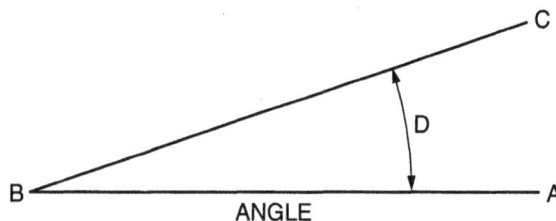

102F04.EPS

Figure 4 ◆ Angle.

There are several kinds of angles:

- *Right angle* – This angle has rays that are perpendicular to one another (*Figure 5*). The measure of this angle is always 90°.
- *Straight angle* – This angle does not look like an angle at all. The rays of a straight angle lie in a straight line, and the angle measures 180°.
- *Acute angle* – An angle less than 90°.
- *Obtuse angle* – An angle greater than 90°, but less than 180°.
- *Adjacent angles* – When three or more rays meet at the same vertex, the angles formed are said to be adjacent (next to) one another. In *Figure 6*, the angles ∠ABC and ∠CBD are adjacent angles. The ray BC is said to be common to both angles.

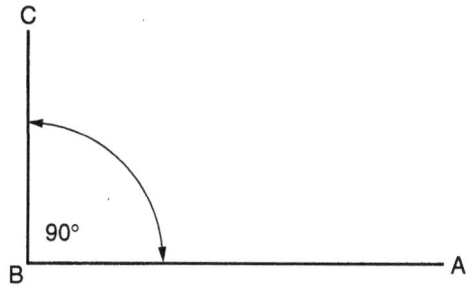

RIGHT ANGLE

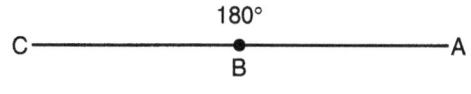

STRAIGHT ANGLE

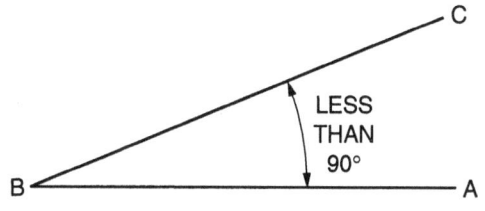

ACUTE ANGLE

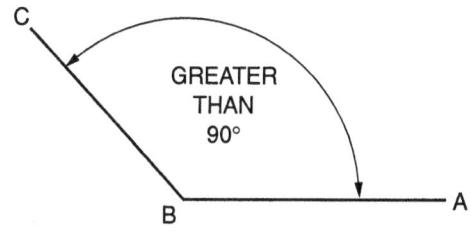

OBTUSE ANGLE

102F05.EPS

Figure 5 ◆ Right, straight, acute, and obtuse angles.

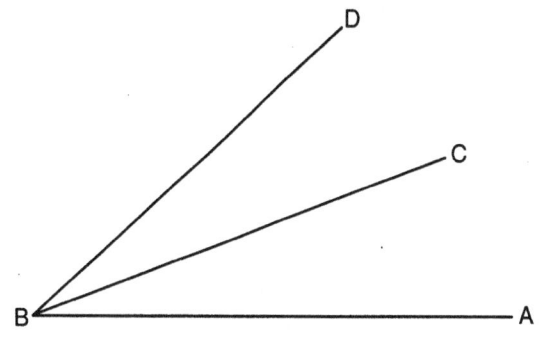

ADJACENT ANGLES

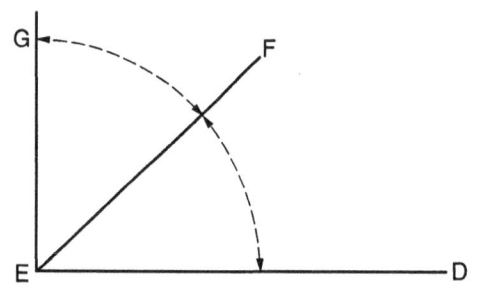

COMPLEMENTARY ANGLES

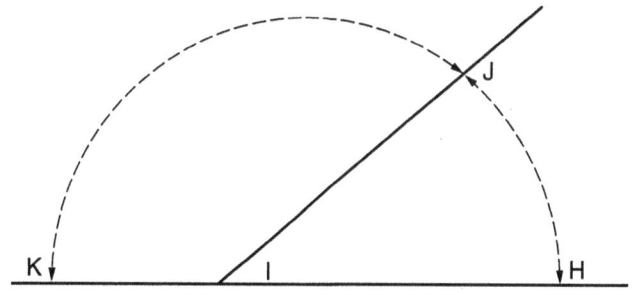

SUPPLEMENTARY ANGLES

102F06.EPS

Figure 6 ◆ Adjacent, complementary, and supplementary angles.

- *Complementary angles* – Two adjacent angles that have a combined total measure of 90°. In *Figure 6*, ∠DEF is complementary to ∠FEG.
- *Supplementary angles* – Two adjacent angles that have a combined total measure of 180°. In *Figure 6*, ∠HIJ is supplementary to ∠JIK.

3.1.4 Polygons

A polygon is formed when three or more straight lines are joined in a regular pattern. Some of the most familiar polygons are shown in *Figure 7*.

As shown, they have common names which generally refer to their number of sides. When all sides of a polygon have equal length and all internal angles are equal, it is called a regular polygon.

Each of the boundary lines forming the polygon is called a side of the polygon. The point at which any two sides of a polygon meet is called a vertex of the polygon. The perimeter of any polygon is equal to the sum of the lengths of each of the sides.

The sum of the interior angles of any polygon is equal to:

$(n - 2) \times 180°$

Where:

n = the number of sides

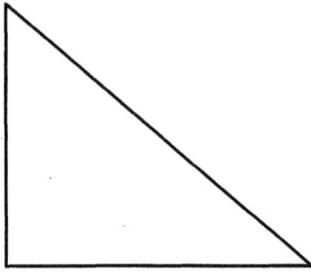 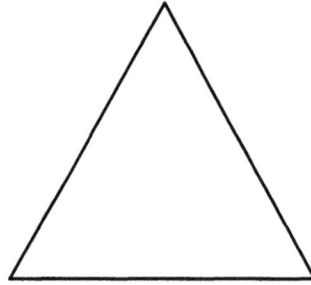 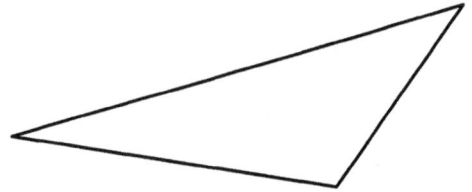

| RIGHT TRIANGLE | EQUILATERAL TRIANGLE | SCALENE TRIANGLE |

TRIANGLES

 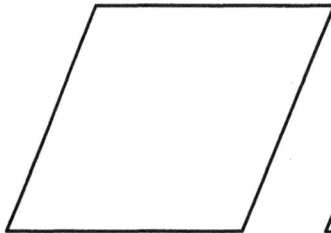 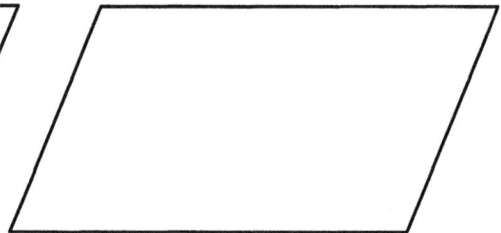

| SQUARE | RHOMBUS | PARALLELOGRAM |

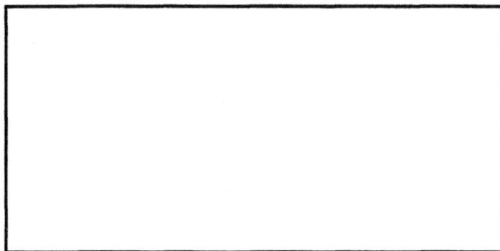 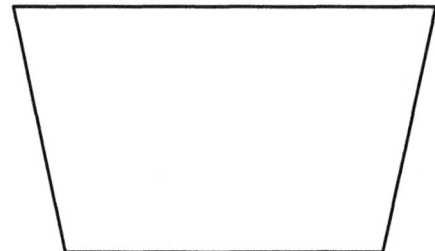

| RECTANGLE | TRAPEZOID |

QUADRILATERALS

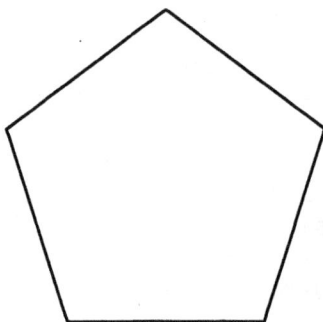 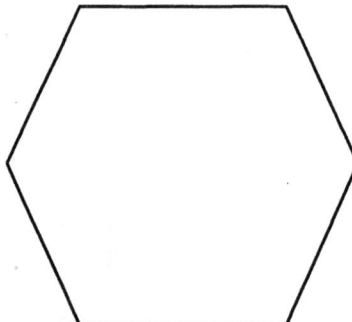 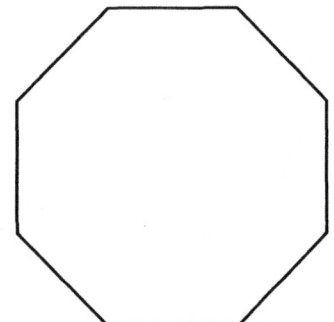

| PENTAGON | HEXAGON | OCTAGON |

REGULAR POLYGONS

102F07.EPS

Figure 7 ◆ Common polygons.

For example, the sum of the interior angles for a square is 360°:

$(4-2) \times 180° = 360°$

For a triangle, it is 180°:

$(3-2) \times 180° = 180°$

3.1.5 Triangles

As mentioned previously, triangles are three-sided polygons. *Figure 8* shows three different types of triangles. A regular polygon with three equal sides is called an equilateral triangle. Two types of irregular triangles are the isosceles (having two sides of equal length) and the scalene (having all sides of unequal length). An important fact to remember about triangles is that the sum of the three angles of any triangle equals 180°. As shown, all three sides of an equilateral triangle are equal. In such a triangle, the three angles are also equal. The isosceles triangle has two equal sides, with the angles opposite the equal sides also being equal to each other.

Triangles are also classified according to their interior angles (*Figure 9*). If one of the three interior angles is 90°, the triangle is called a right triangle. If one of the three interior angles is greater than 90°, the triangle is called an obtuse triangle. If each of the interior angles is less than 90°, the triangle is called an acute triangle. The sum of the three interior angles of any triangle is always 180°.

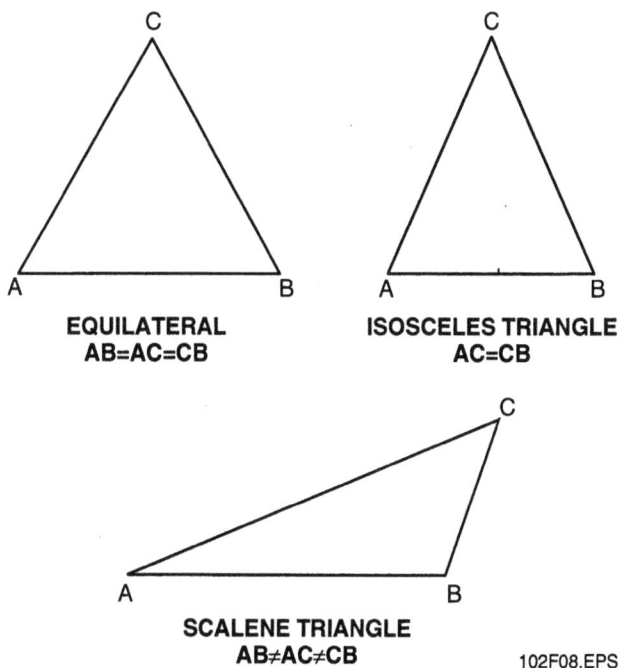

Figure 9 ◆ Right, acute, and obtuse triangles.

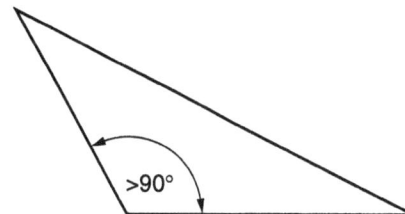

This is helpful to remember whenever you know two angles of a triangle and need to calculate the third. For example, if two angles of a triangle are 28° and 42°, the unknown angle equals 110° ($180° - 42° - 28° = 110°$).

3.2.0 Working with Right Triangles

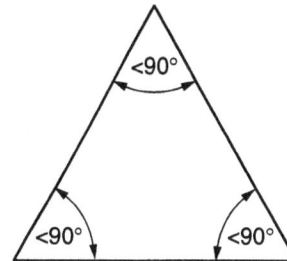

Perhaps the most used shape in construction is the right triangle. Any vertical object or structure, such as a telephone pole or the face of a building, is part of a right triangle. If you draw an imaginary line from a point on the ground to the top of the structure, such as the top of a pole or the roof of a building, that line forms the **hypotenuse** of a right triangle. The base of the triangle extends from the bottom of the pole or structure to the point on the ground that was the starting point for your line.

Because the right triangle has one right angle, the other two angles are acute angles. They are also complementary angles, the sum of which

Figure 8 ◆ Triangles.

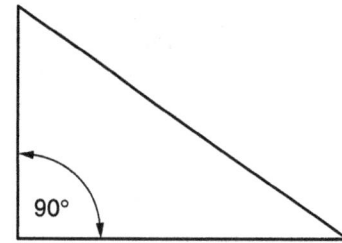

equals 90°. The right triangle has two sides perpendicular to each other, thus forming the right angle. To aid in writing equations, the sides and angles of a right triangle are labeled as shown in *Figure 10*. Normally, capital (uppercase) letters are used to label the angles and lowercase letters are used to label the sides. The third side, which is always opposite the right angle (C), is called the hypotenuse. It is always longer than either of the other two sides. The other sides can be remembered as a, for altitude, and b, for base. Note that the letters that label the sides and angles are opposite each other. For example, side a is opposite angle A, and so forth.

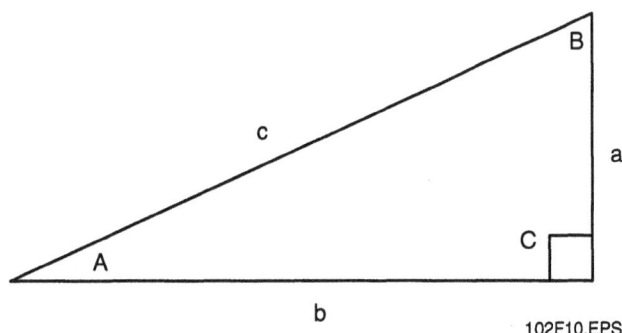

Figure 10 ◆ Common labeling of angles and sides in a right triangle.

3.2.1 Right Triangle Calculations Using the Pythagorean Theorem

If you know the length of any two sides of a right triangle, you can calculate the length of the third side using a rule called the Pythagorean theorem. It states that the square of the hypotenuse (c) is equal to the sum of the squares of the remaining two sides (a and b). Expressed mathematically:

$$c^2 = a^2 + b^2$$

You may rearrange to solve for the unknown side as follows:

$$a = \sqrt{c^2 - b^2}$$
$$b = \sqrt{c^2 - a^2}$$
$$c = \sqrt{a^2 + b^2}$$

For example, assume you had a right triangle with an altitude (side a) equal to 8' and a base (side b) equal to 12'. To find the length of the hypotenuse (side c), proceed as follows:

$$c = \sqrt{a^2 + b^2}$$
$$c = \sqrt{(8')^2 + (12')^2}$$
$$c = \sqrt{64 + 144}$$
$$c = \sqrt{208}$$
$$c = 14.422'$$

To determine the actual length of the hypotenuse using the formula above, it is necessary to calculate the square root of the sum of the sides squared. Fortunately, this is easy to do using a scientific calculator. On many calculators, you simply key in the number and press the square root [√] key. On some calculators, the square root does not have a separate key. Instead, the square root function is the inverse of the [x^2] key, so you have to press [INV] or [2nd F], depending on your calculator, followed by [x^2], to obtain the square root.

To see a practical application of the Pythagorean theorem, let's say you know the length of a support cable attached to the top of a telephone pole. You wish to know the height of the pole, but measuring it from the ground would not be practical. You can, however, easily measure the distance between the end of the cable attached to the ground and the bottom of the pole—which forms the base of a right triangle.

If the length of the cable (the hypotenuse, or c from the equation) is 25' and the distance from the cable to the base of the pole is 10', we know the c and b parts of formula:

$$c^2 = a^2 + b^2$$
$$25^2 = a^2 + 10^2$$
$$625 = a^2 + 100$$

Subtracting 100 from both sides of the equation, we get the following:

$$525 = a^2$$

The square root of 525, therefore, is the height of the pole (roughly 22.9').

3.2.2 The 3-4-5 Rule

The 3-4-5 rule describes a simple method for laying out or checking 90° angles (right angles) and requires only the use of a tape measure. The rule is based on the Pythagorean theorem and has been used in building construction for centuries. The numbers 3-4-5 represent dimensions in feet that describe the sides of a right triangle. Right triangles that are multiples of the 3-4-5 triangle are commonly used, such as 9-12-15, 12-16-20, 15-20-25, and so forth. The specific multiple used is determined by the relative distances involved in the job being laid out or checked.

An example of the 3-4-5 rule using the multiples 15-20-25 is shown in *Figure 11*. In order to square or check a corner as shown in the example, first measure and mark 15'-0" down the line in one direction, then measure and mark 20'-0" down the line in the other direction. The distance measured between the 15'-0" and 20'-0" points must be exactly 25'-0" to ensure that the angle is a perfect right angle.

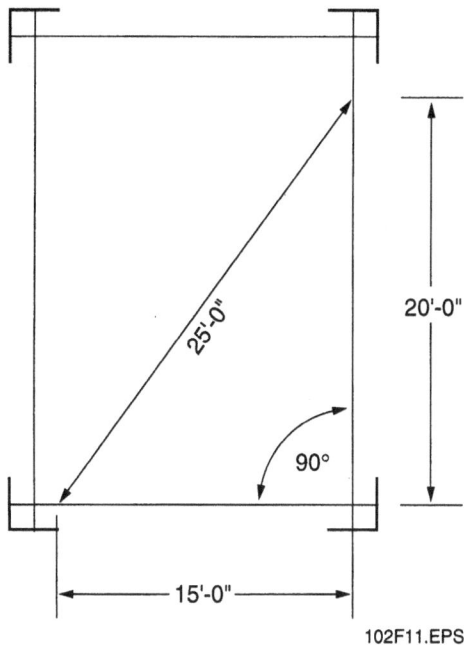

Figure 11 ◆ The 3-4-5 rule.

4.0.0 ◆ CONVERSION PROCESSES

In *Core Curriculum*, you learned basic measurements, conversion processes, and the basics of the metric system. When performing surveys you will need to convert between measurement systems and perform calculations based on the converted figures, so these concepts are discussed in greater depth in this module. You will also learn how to apply conversion processes to area and volume computations.

4.1.0 Linear Measure

In the English system of measurement, linear measure is expressed in inches and feet, which are further divided into fractional or decimal parts. You know that there are 12 inches in 1 foot and that there are 36 inches or 3 feet in 1 yard. If you want to convert inches to yards, you must first divide the number of inches by 12 and then divide that number by 3, or you can divide the number of inches by 36. Study the following example:

How many yards are in 72 inches?

72 inches ÷ 12 inches = 6 inches ÷ 3 inches = 2 yards

or

72 inches ÷ 36 inches = 2 yards

In the metric system, linear measure is expressed in meters, which are further divided into centimeters and millimeters. The metric system uses the number 10 as the basis for calculations. Although the United States has not completely converted to the metric system, many products, tools, and drawings

in the U.S. and Canada are measured or expressed using the metric system. Some job specifications also require the use of this system, so using the metric system and learning how to convert from one system to the other will help you in your career.

Because the metric system is based on 10, calculations are fairly easy. There are 100 centimeters (cm) in a meter (m) and 1,000 millimeters (mm) in a meter. One meter is equal to 39.37 inches. How many inches are there in one centimeter? To answer this question, divide 39.37 by 100. One centimeter equals 0.3937 inches (or 0.394 inches rounded off). To find how many inches are in one millimeter, divide 39.37 by 1,000. One millimeter equals 0.03937 inches (or 0.0394 inches rounded off). Notice that to solve each problem, you can simply move the decimal point to the left: two places when you divide by 100 and three places when you divide by 1,000.

How many centimeters are in one inch? You know that one centimeter equals 0.3937 inches, so you divide as follows:

1 centimeter ÷ 0.3937 inches = 2.54 centimeters

To find how many millimeters are in one inch, you divide by 0.03937:

1 millimeter ÷ 0.03937 inches = 25.4 millimeters

The answers to these problems—2.54 centimeters and 25.4 millimeters—are called conversion factors. To convert an English measure to a metric measure, multiply the English measure by the appropriate metric conversion factor. A list of common conversion factors are contained in the *Appendix* at the back of this module. Notice that when the problem involves a fraction, you must convert the fraction to a decimal and then multiply. Study the following examples:

5 inches = 12.7 centimeters (5 × 2.54)

5 inches = 127 millimeters (5 × 25.4)

¾ inches = 1.905 centimeters (0.75 × 2.54)

¾ inches = 19.05 millimeters (0.75 × 25.4)

4.2.0 Converting Between Measurement Systems

When using formulas to calculate numerical values for lengths, distances, and angles, the answers obtained either by hand or by calculator are normally expressed as a decimal. When using surveying instruments, the readouts for angular measurements are typically expressed in degrees, minutes, and seconds. Construction drawings and plot plans often express similar measurements differently. For this reason, it may be necessary to convert between different measurement systems.

This section explains how to make conversions between measurement systems commonly used in construction. Conversion tables available in many trade-related reference books can be used for this purpose. *Table 1* is an example of a conversion table. However, if conversion tables are not readily available, you should be familiar with the methods for making such conversions mathematically.

4.2.1 Converting Decimal Feet to Feet and Inches

To convert values given in decimal feet into equivalent feet and inches, use the following procedure. For our example, we will convert 45.3646 feet to feet and inches (rounded to the nearest ⅛ inches).

Step 1 Subtract 45' from 45.3646' = 0.3646'.

Step 2 Convert 0.3646' to inches by multiplying 0.3646' by 12 = 4.3752".

Step 3 Subtract 4" from 4.3752" = 0.3752".

Step 4 Convert 0.3752" into eighths of an inch by multiplying 0.3752" by 8 = 3.0016 eighths or, when rounded off, ⅜. Therefore, 45.3646' = 45'-4⅜".

4.2.2 Converting Feet and Inches to Decimal Feet

To convert values given in feet and inches (and inch-fractions) into equivalent decimal feet values, use the following procedure. For our example, we will convert 45'-4⅜" to decimal feet:

Step 1 Convert the inch-fraction ⅜" to a decimal. This is done by dividing the numerator of the fraction (top number) by the denominator of the fraction (bottom number). For example, ⅜" = 0.375.

Step 2 Add the 0.375" to 4" to obtain 4.375".

Step 3 Divide 4.375" by 12 to obtain 0.3646'.

Step 4 Add 0.3646' to 45' to obtain 45.3646'. Therefore, 45'-4⅜" = 45.3646'.

4.3.0 Converting Area Measurements

In *Core Curriculum*, you learned to calculate the areas of basic shapes. As you know, the area of any rectangle is equal to the length multiplied by the width. In the familiar inch-pound system, the units of area are square feet (ft^2) or square inches (in^2). *Figure 12* shows the application of this concept.

The area of a circle is found using the following formula:

$$Area = \pi r^2$$

Where:

A = the area of the circle

π = a constant of 3.14159 (often rounded off to 3.14)

r = the radius (distance from the center to the edge of the circle)

Figure 13 shows the application of this concept.

When converting areas from one measurement system to the other, every dimension must be converted. To convert the dimensions of the square shown in *Figure 12A*, we must convert both the length and the width, as shown in *Figure 14*.

To convert the dimensions of the rectangle shown in *Figure 12B*, we must convert both the length and the width, as shown in *Figure 15*.

When converting the dimensions of a circle, only the radius has a measured dimension that must be converted. In the example shown in *Figure 16*, centimeters must be converted into inches.

Table 1 Inches into Decimal Feet Conversion Table

Inches	Decimal Equivalent	Inches	Decimal Equivalent
¹⁄₁₆	0.0052	1	0.0833
³⁄₃₂	0.0078	2	0.1667
⅛	0.0104	3	0.2500
³⁄₁₆	0.0156	4	0.3333
¼	0.0208	5	0.4167
⁵⁄₁₆	0.0260	6	0.5000
⅜	0.0313	7	0.5833
½	0.0417	8	0.6667
⅝	0.0521	9	0.7500
¾	0.0625	10	0.8333
⅞	0.0729	11	0.9167

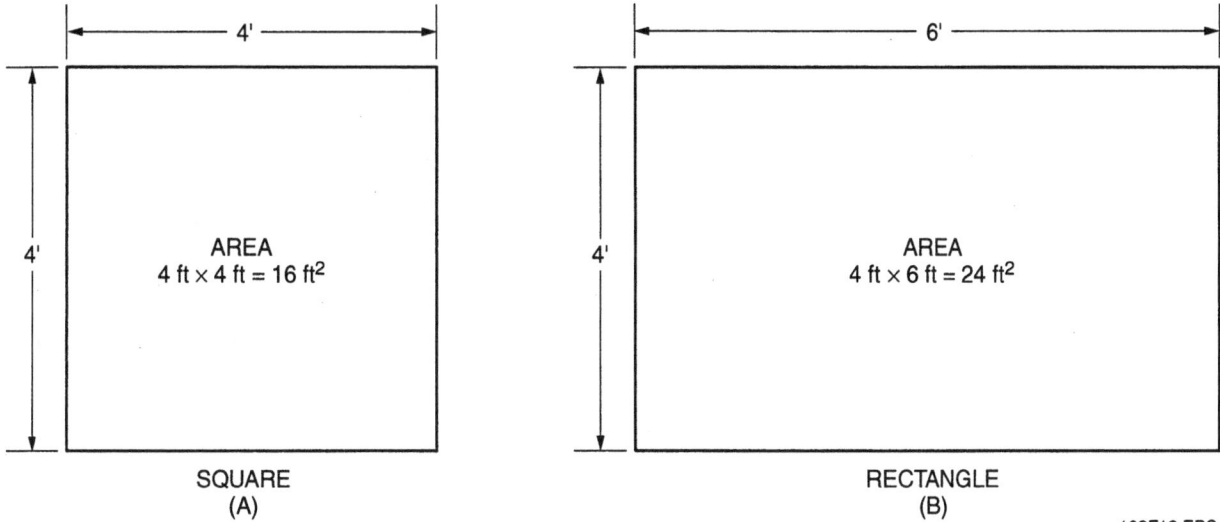

Figure 12 ◆ Measuring the area of a square and rectangle.

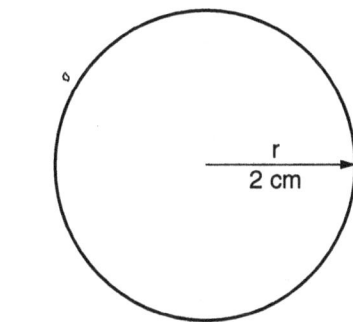

Figure 13 ◆ Measuring the area of a circle.

4.4.0 Volume

In *Core Curriculum*, you learned to calculate the volumes of basic shapes. As you know, the volume of a square or rectangle is the product of three lengths. It has units of length × length × length, or length cubed. In the familiar inch-pound system, the most common units of volume are cubic feet (ft^3) or cubic inches (in^3).

Figure 17 shows the three-dimensional measurement of a rectangle. Its volume is calculated by multiplying length × width × height.

This will suffice for finding the volume of box-shaped containers. The next shape we will discuss is that of a tank or other cylindrical object, as shown in *Figure 18*.

The volume of the box can be considered the area of any side times the height. Likewise, the volume of a cylindrical tank can be simplified by considering it as the area of the circle times its depth (or height).

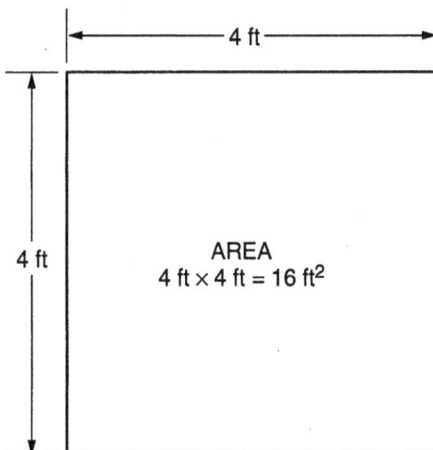

Figure 14 ◆ Conversion of units for a square.

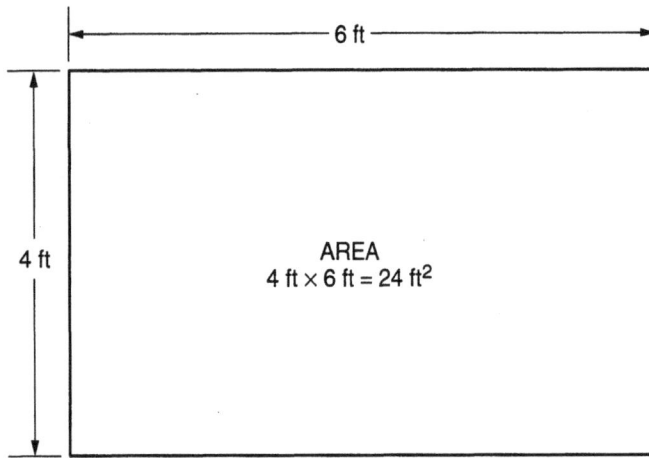

$$\text{AREA} = \text{LENGTH} \times \text{WIDTH}$$

$$= 4\ \text{ft} \times 6\ \text{ft}$$

$$= (4\ \text{ft})\left(\frac{1\ \text{m}}{3.281\ \text{ft}}\right) \times (6\ \text{ft})\left(\frac{1\ \text{m}}{3.281\ \text{ft}}\right)$$

$$= (1.219\ \text{m}) \times (1.829\ \text{m})$$

$$= 2.229\ \text{m}^2$$

102F15.EPS

Figure 15 ◆ Conversion of units for a rectangle.

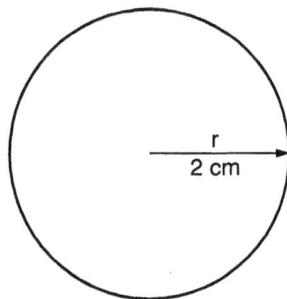

$$\text{AREA} = \pi r^2$$

$$= (3.14)\ [(2\ \text{cm})\ (0.3937\ \text{in/cm})]\ [(2\ \text{cm})\ (0.3937\ \text{in/cm})]$$

$$= (3.14)\ (0.7874\ \text{in})\ (0.7874\ \text{in})$$

$$= (3.14)\ (0.6200\ \text{in})$$

$$= 1.95\ \text{in}^2$$

102F16.EPS

Figure 16 ◆ Conversion of units for a circle.

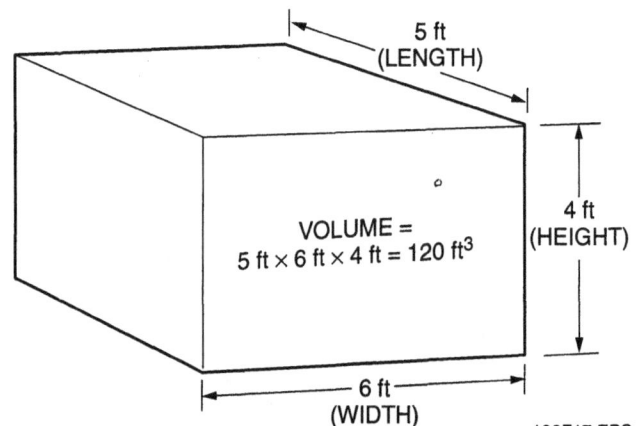

102F17.EPS

Figure 17 ◆ Volume of a rectangle.

When converting these two volumes from one measurement system to the other, remember that every dimension must be independently converted. For converting the dimensions of the box, we must convert the length, width, and height, as shown in *Figure 19*.

In the example of the cylindrical tank, the same rule applies, as shown in *Figure 20*.

4.4.1 Applying Volume Measurements: Estimating Concrete Volume

One possible application of volume measurements and calculations with which you should be familiar is estimating concrete volume. Fortunately, most concrete structures can be divided into rectangular or circular shapes, individually estimated,

and the results added together to obtain the required volume of concrete. For instance, a floor is a rectangular horizontal slab. A footing and a wall can be divided into a long rectangular shape representing the footing, and a vertical slab representing the wall. The volume of each can be calculated and the results added together for the total volume. Blueprints for a project will provide the dimensions for the various portions of a concrete structure, and these dimensions are used to calculate the volume of concrete required for the structure.

A number of methods can be used to determine the volume of concrete required for a rectangular object. One method uses the following formula:

Cubic yards of concrete (rounded up to next ¼ yard) =

$$\frac{\text{width or height (ft)} \times \text{length (ft)} \times \text{thickness (ft)}}{27\ (\text{ft}^3/\text{yd})}$$

To use the formula, all dimensions in inches must be converted to feet and/or fractions of a foot and then into a decimal equivalent.

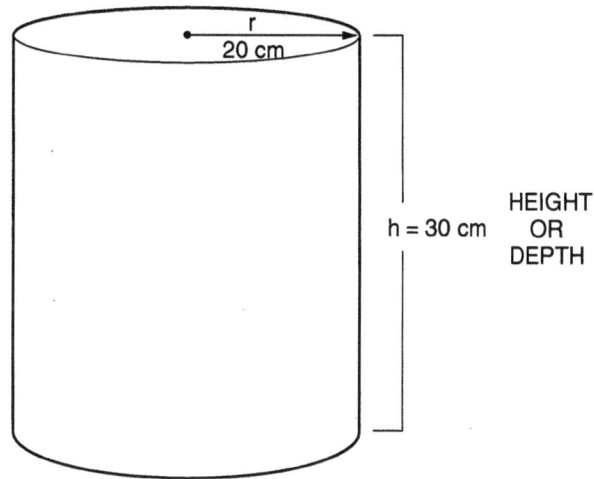

VOLUME $= (\pi r^2) \times h$ or AREA OF THE CIRCLE $\times$ HEIGHT

$$\overbrace{\hspace{3cm}}^{\text{AREA}} \overbrace{\hspace{1.5cm}}^{\text{HEIGHT}}$$

$= \{(3.14)\,[(20\text{ cm})\,(20\text{ cm})]\} \times (30\text{ cm})$

$= [(3.14)\,(400\text{ cm}^2)] \times (30\text{ cm})$

$= (1{,}256\text{ cm}^2) \times (30\text{ cm})$

$= 37{,}680\text{ cm}^3$

102F18.EPS

Figure 18 ◆ Volume of a cylindrical tank.

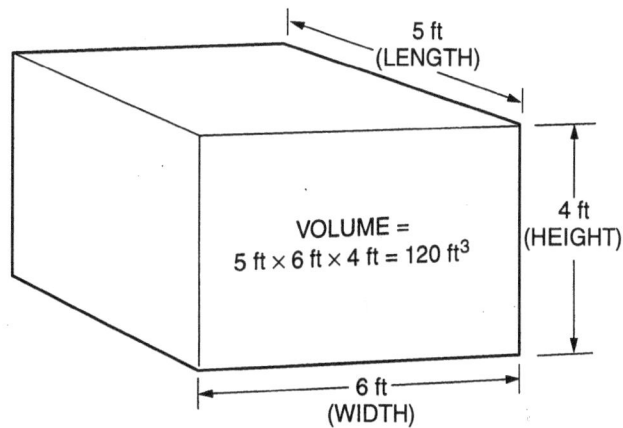

VOLUME = LENGTH $\times$ WIDTH $\times$ HEIGHT

$$= (5\text{ ft})\left(\frac{1\text{ m}}{3.281\text{ ft}}\right) \times (6\text{ ft})\left(\frac{1\text{ m}}{3.281\text{ ft}}\right) \times (4\text{ ft})\left(\frac{1\text{ m}}{3.281\text{ ft}}\right)$$

$$= \frac{5\text{ m}}{3.281} \times \frac{6\text{ m}}{3.281} \times \frac{4\text{ m}}{3.281}$$

$= (1.524\text{ m}) \times (1.829\text{ m}) \times (1.219\text{ m})$

$= 3.398\text{ m}^3$

102F19.EPS

Figure 19 ◆ Conversion of units for a box.

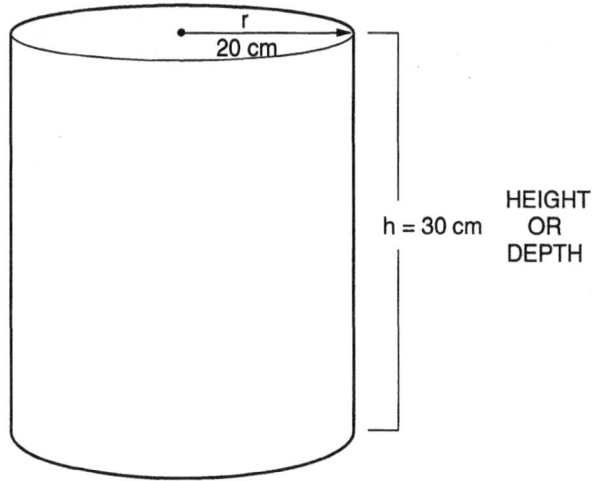

$$\text{VOLUME} = (\pi r^2) \times h$$

$$= (3.14)\,[(20\ cm)\,(0.3937\ in/cm)]\,[(20\ cm)\,(0.3937\ in/cm)]\,[(30\ cm)\,(0.3937\ in/cm)]$$

$$= (3.14)\,(7.874\ in)\,(7.874\ in)\,(11.811\ in)$$

$$= 2{,}299\ in^3$$

102F20.EPS

Figure 20 ◆ Conversion of units for a cylindrical tank.

For example:

7 inches = $\frac{7}{12}$ foot or 7 ÷ 12 = 0.58 foot

8 inches = $\frac{8}{12}$ foot or $\frac{2}{3}$ foot or 2 ÷ 3 = 0.66 foot

23 inches = $\frac{23}{12}$ feet or 23 ÷ 12 = 1.92 feet

The other two methods involve knowing the width or height and length of an area in feet, along with the thickness in inches, and then using a concrete calculator (*Figure 21*) or a concrete table (*Figure 22*) and a simple formula to determine the volume.

The concrete calculator cannot be used when the volume of concrete required is less than one-half yard. Also, the results of the calculator must be interpolated (approximated) by sight between rule marks, which makes the results subject to some variation.

Using the calculator requires the following:

- Locating the thickness in inches of the rectangular object on Scale A and the width (or height) in feet on Scale B.

- Moving the sliding part of the calculator to align the desired thickness value on Scale A with the desired width (or height) value on Scale B.

- Locating the length of the object on Scale C and reading the required concrete volume in cubic yards on Scale D directly opposite the length on Scale C.

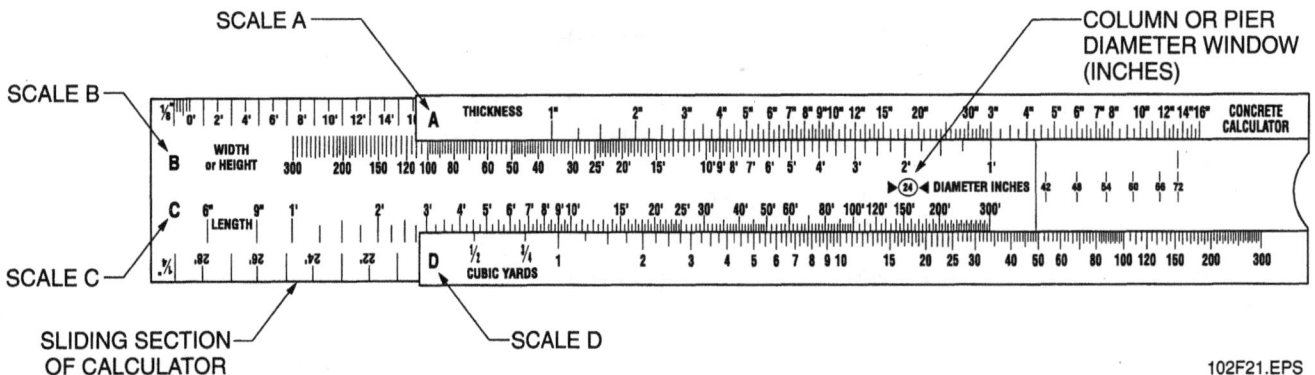

Figure 21 ◆ Typical concrete calculator.

ONE CUBIC YARD OF CONCRETE WILL PLACE:					
THICKNESS	SQ FT	THICKNESS	SQ FT	THICKNESS	SQ FT
1"	324	5"	65	9"	36
1¼"	259	5¼"	62	9¼"	35
1½"	216	5½"	59	9½"	34
1¾"	185	5¾"	56	9¾"	33
2"	162	6"	54	10"	32.5
2¼"	144	6¼"	52	10¼"	31.5
2½"	130	6½"	50	10½"	31
2¾"	118	6¾"	48	10¾"	30
3"	108	7"	46	11"	29.5
3¼"	100	7¼"	45	11¼"	29
3½"	93	7½"	43	11½"	28
3¾"	86	7¾"	42	11¾"	27.5
4"	81	8"	40	12"	27
4¼"	76	8¼"	39	15"	21.5
4½"	72	8½"	38	18"	18
4¾"	68	8¾"	37	24"	13.5

102F22.EPS

Figure 22 ◆ Portion of a typical concrete table.

The concrete table requires the following:

- Converting any dimensions in inches to feet and/or fractions of a foot and then into a decimal equivalent, then calculating the area:

  Area (square foot) = length (foot) × width or height (foot)

- Locating the desired thickness (in inches) in the table and noting the square foot number in the table for that thickness. The square foot number represents the area that one cubic yard of concrete will cover at that thickness.
- Dividing the calculated area by the square foot number from the table to determine the cubic yards of concrete required.

  Cubic yards required (rounded up to next ¼ yard) =

  $$\frac{\text{area (square foot)}}{\text{square foot (number from table)}}$$

Using the concrete table (*Figure 22*), determine the amount of concrete required for the partial wall, footing, and floor slab plan shown in *Figure 23*.

Step 1 The entire footing and wall length must be determined. Because the wall is centered on the footing, the wall length is the same as the footing length:

Footing/wall length = 20' + (15' − 2') = 33'

Step 2 Determine the area of the footing and wall:

Wall area = 3' × 33' = 99 square feet

Footing area = 2' × 33' = 66 square feet

Step 3 Determine the floor slab length, width, and area:

Length = 20' − 16" = 20' − 1.33' = 18.67'

Width = 15' − 16" = 15' − 1.33' = 13.67'

Slab area = 18.67' × 13.67' = 255.2 square feet

Step 4 From the concrete table in *Figure 23*, determine the square foot number for the thicknesses of the wall, footing, and slab:

Wall thickness = 8 inches = 40 square feet

Footing thickness = 8 inches = 40 square feet

Slab thickness = 6 inches = 54 square feet

Step 5 Determine the cubic yards of concrete required by using the formula:

Cubic yards required =

$$\frac{\text{area (square foot)}}{\text{square foot (number from table)}}$$

Wall = $\frac{99}{40}$ = 2.48 cubic yards

Footing = $\frac{66}{40}$ = 1.65 cubic yards

Slab = $\frac{255.2}{54}$ = 4.73 cubic yards

Step 6 Add the wall, footing, and slab volumes:

2.48 + 1.65 + 4.73 = 8.86
(rounded up to 9 cubic yards)

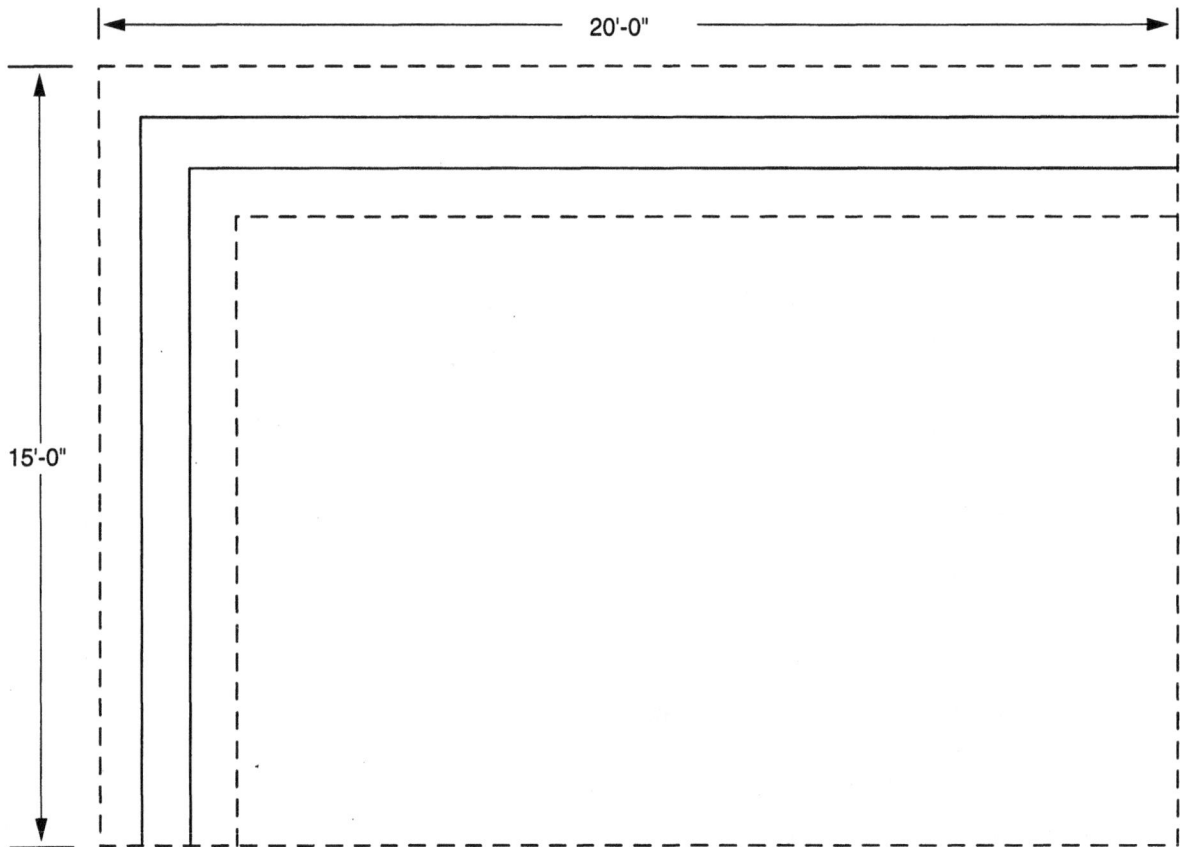

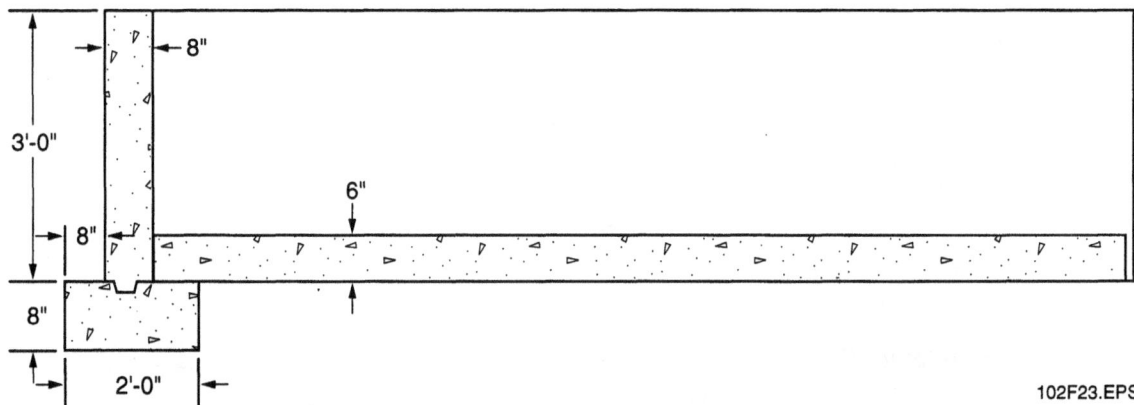

Figure 23 ◆ Floor plan.

102F23.EPS

For the plan shown, 9 cubic yards of concrete will be required.

4.4.2 Applying Volume Measurements: Wet Measurements

In our previous discussions of volume measurements, we used the shape of the container to calculate the volume. This is referred to as a dry measure. In practice, we are much more used to dealing with wet measures, such as the amount of a fluid that would fill the volume. Common wet measures in the inch-pound system include the pint, quart, and gallon. The metric system also uses wet measuring units. The liter is the most common and is about 5% greater than a quart.

By definition, one liter is one cubic decimeter. In other words, a cube with each side equal to one decimeter (or ten centimeters) will hold one liter of fluid. Knowing the wet measures for a substance allows easy handling and measuring of fluids because the fluid will conform to the shape of the container. If you had to recalculate the amount each time you moved a fluid, you would soon see the advantage of using wet measures.

Table 2 shows the volume relationships between the liter and the dry volume of the cubic meter in the metric system and the pint and gallon in the inch-pound system.

Figure 24 shows that the same metric prefixes that apply to the meter can be used with the liter.

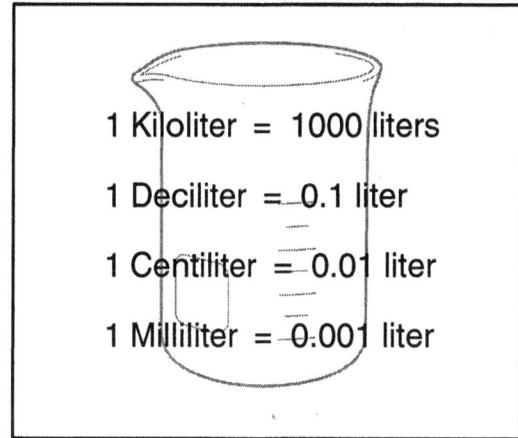

102F24.EPS

Figure 24 ◆ Common metric prefixes used with volumes.

Summary

Surveyors and site layout technicians work with measured data, drawings and site plans, geometric shapes, distances, areas, and volumes in analyzing and documenting the layout of a given work site. As a site layout technician, you will apply mathematical formulas constantly while on the job. You must understand the formulas as well as how to convert between measurement systems and units in order to apply your measurements and figures properly.

Table 2 Volume Relationships

Unit	Cubic Meter	Gallon	Liter	Pint
Cubic meter (m³)	1	0.0375	0.001	0.0004732
U.S. gallon (gal)	264.2	1	0.242	0.125
Liter (L)	1,000	3.785	1	0.4732
U.S. pint (pt)	2,113	8	2.113	1

Review Questions

1. The area of a rectangle is represented by the formula _____.
   a. area equals length times width
   b. area equals height times radius
   c. area equals pi times radius squared
   d. area equals length times width times height

2. In the formula area equals pi times r squared, r (the radius) is _____.
   a. a variable
   b. a coefficient
   c. a constant
   d. 3.14

3. The order of operations for solving a complicated equation is _____.
   a. subtract, add, divide, multiply
   b. add, multiply, subtract, divide
   c. divide, add, subtract, multiply
   d. multiply, divide, add, subtract

4. The correct answer for the equation $3 \times 4 + 2 \times 6$ is _____.
   a. 24
   b. 48
   c. 84
   d. 144

5. The divisor that is multiplied by itself to produce a number is the _____ of that number.
   a. square
   b. square root
   c. multiplier
   d. dividend

6. The square of 36 is _____.
   a. 6
   b. 72
   c. 216
   d. 1,296

7. A number that has a whole number as its square root is called a _____.
   a. cube
   b. perfect square
   c. pure square
   d. divisor

8. The value of 144^2 is equal to _____.
   a. 2,073
   b. 20,736
   c. 207,360
   d. 2,073,600

9. The square root of 162,500 is _____.
   a. 4.03113
   b. 40.3113
   c. 403.113
   d. 4,031.13

10. Area equals pi times radius squared is the formula for the area of a _____.
    a. square
    b. rectangle
    c. circle
    d. triangle

11. The value of pi is approximately _____.
    a. 3
    b. 3.4
    c. 3.14
    d. 3.21

12. Every site at which you work can be divided into _____.
    a. loading zones
    b. geometric shapes
    c. areas of responsibility
    d. legal zones

13. Each geometric point represents _____.
    a. length and width
    b. height
    c. position
    d. area

14. Defining a line requires a minimum of ____ points.
    a. 1
    b. 2
    c. 3
    d. 4

15. A line that forms a right angle with another line is _____ that line.
    a. parallel to
    b. perpendicular to
    c. congruent to
    d. in series with

16. In a circle, the shortest distance from any point on the curve through the center to a point directly opposite is _____.
    a. pi
    b. the radius
    c. the diameter
    d. the circumference

17. A circle is divided into _____ degrees.

a. 90
b. 180
c. 270
d. 360

18. One degree is equal to _____ minutes.

a. 1
b. 30
c. 60
d. 3,600

19. When working with a circle, the quantity πd represents _____.

a. circumference
b. radius
c. diameter
d. cord

20. An angle is made up of two straight lines that meet at a _____.

a. radius
b. ray
c. flat
d. vertex

21. An angle less than 90 degrees is called a(n) _____ angle.

a. acute
b. right
c. obtuse
d. adjacent

22. Adjacent angles that have a combined measure of 90 degrees are called _____ angles.

a. supplementary
b. complementary
c. obtuse
d. right

23. Each of the boundary lines forming a polygon is called a(n) _____.

a. side
b. vertex
c. flat
d. radius

24. The sum of the interior angles for a triangle is _____ degrees.

a. 90
b. 180
c. 270
d. 360

25. Two angles of a triangle are 90 degrees and 60 degrees. The remaining angle must be _____ degrees.

a. 30
b. 60
c. 90
d. 180

26. A triangle in which all sides are of unequal length is called a(n) _____ triangle.

a. equilateral
b. isosceles
c. right
d. scalene

27. The two complementary angles in a right triangle are _____.

a. acute
b. obtuse
c. hypotenuses
d. parallel

28. The two sides of a right triangle measure 4' and 8'. The hypotenuse must therefore measure _____.

a. 4.94'
b. 8.94'
c. 9.94'
d. 10.94

29. Each of the following is a form of the Pythagorean theorem *except* _____.

a. $c^2 = a^2 + b^2$
b. $a = \sqrt{c^2 - b^2}$
c. $b = \sqrt{c^2 - a^2}$
d. $c = \sqrt{a^2 - b^2}$

30. The 3-4-5 rule is a simple method for laying out or checking _____.

a. circular area
b. right angles
c. nonlinear calculations
d. volume

31. How many yards are in 72 inches?

a. One
b. Two
c. Three
d. Four

32. 7.375 inches is equal to _____ millimeters.

a. 18.733
b. 187.325
c. 1,873.25
d. 18,732.50

33. Convert 6.875 feet to feet and inches.
    a. 6'-2¼"
    b. 6'-8¼"
    c. 6'-10½"
    d. 6'-11½"

34. Convert 20'-3⅛" into decimal feet.
    a. 20.2604
    b. 20.6204
    c. 20.6206
    d. 20.9402

35. When converting areas from one measurement system to another, _____.
    a. convert every dimension and then calculate area
    b. calculate area and then convert the result
    c. convert one dimension and then convert the result
    d. consult a conversion table for the area figure

36. A room has a length of 20 feet and a width of 10 feet. The area of this room in square meters is _____.
    a. 18.58 m²
    b. 56.28 m²
    c. 200 m²
    d. 2,152.99 m²

37. What is the volume of concrete rounded up to the next ¼ yard needed for a 20-foot-tall concrete column having a diameter of 3 feet?
    a. 3¼ cubic yards
    b. 5¼ cubic yards
    c. 5½ cubic yards
    d. 6¼ cubic yards

38. The pint, quart, and gallon are common _____.
    a. dry measurements
    b. wet measurements
    c. area measurements
    d. metric measures

39. A cubic decimeter is _____ liter(s).
    a. 1
    b. 10
    c. 100
    d. 1,000

40. How many U.S. gallons are in one cubic meter?
    a. 3.785
    b. 125.4
    c. 264.2
    d. 2,113

Trade Terms Introduced in This Module

Acute angle: Any angle less than 90 degrees.

Complementary angles: Any two angles that add up to 90 degrees.

Formula: A concise statement that uses symbols or letters to describe the relationship between certain values, such as line segment lengths and angles.

Line segment: A straight line that has a starting point and an ending point.

Hypotenuse: The slanted line segment that is directly across from the right angle in a right triangle. The longest side of a right triangle.

Pythagorean theorem: A formula stating that in a right triangle, the square of the length of the hypotenuse equals the sum of the squares of the lengths of the other two segments ($c^2 = a^2 + b^2$).

Polygon: Any enclosed figure connected by three or more straight line segments.

Right triangle: A polygon having three sides, one 90-degree inside angle, and two additional inside angles that are less than 90 degrees each, but when they are added together they total 90 degrees.

Square: To multiply a number or value times itself. For example, $6 \times 6 = 36$, or 6 squared equals 36. Another way to show this is 6^2.

Square root: The number that when multiplied by itself produces the given number. For example, the square root of 49 is 7 and the square root of 36 is 6. All numbers have a square root, but the only numbers between 1 and 100 that have even square roots are 1, 4, 9, 16, 25, 36, 49, 64, 81, and 100.

Triangle: A polygon having three sides and whose inside angles total 180 degrees.

Conversion Factors

MULTIPLY		BY		TO OBTAIN
LENGTH				
Centimeters	×	.3937	=	Inches
Fathoms	×	6.0	=	Feet
Feet	×	12.0	=	Inches
Feet	×	.3048	=	Meters
Inches	×	2.54	=	Centimeters
Kilometers	×	.6214	=	Miles
Meters	×	3.281	=	Feet
Meters	×	39.37	=	Inches
Meters	×	1.094	=	Yards
Miles	×	5280.0	=	Feet
Miles	×	1.609	=	Kilometers
Rods	×	5.5	=	Yards
Yards	×	.9144	=	Meters
AREA				
Acres	×	43560.0	=	Square feet
Acres	×	4840.0	=	Square yards
Circular mils	×	7.854×10^{-7}	=	Square inches
Circular mils	×	.7854	=	Square mils
Square centimeters	×	.155	=	Square inches
Square feet	×	144.0	=	Square inches
Square feet	×	.0929	=	Square meters
Square inches	×	6.452	=	Square centim.
Square meters	×	1.196	=	Square yards
Square miles	×	640.0	=	Acres
Square mils	×	1.273	=	Circular mils
Square yards	×	.8361	=	Square meters
VOLUME				
Cubic feet	×	.0283	=	Cubic meters
Cubic feet	×	7.481	=	Gallons
Cubic inches	×	.5541	=	Ounces (fluid)
Cubic meters	×	35.31	=	Cubic feet
Cubic meters	×	1.308	=	Cubic yards
Cubic yards	×	.7646	=	Cubic meters
Gallons	×	.1337	=	Cubic feet
Gallons	×	3.785	=	Liters
Liters	×	.2642	=	Gallons
Liters	×	1.057	=	Quarts (liquid)
Ounces (fluid)	×	1.805	=	Cubic inches
Quarts (liquid)	×	.9463	=	Liters

102A01.TIF

CONVERSION FACTORS

MULTIPLY		BY		TO OBTAIN
FORCE AND WEIGHT				
Grams	×	.0353	=	Ounces
Kilograms	×	2.205	=	Pounds
Newtons	×	.2248	=	Pounds (force)
Ounces	×	28.35	=	Grams
Pounds	×	453.6	=	Grams
Pounds (force)	×	4.448	=	Newtons
Tons (short)	×	907.2	=	Kilograms
Tons (short)	×	2000.0	=	Pounds
TORQUE				
Gram-centimeters	×	.0139	=	Ounce-inches
Newton-meters	×	.7376	=	Pound-feet
Newton-meters	×	8.851	=	Pound-inches
Ounce-inches	×	72.0	=	Gram-centimeters
Pound-feet	×	1.3558	=	Newton-meters
pound-inches	×	.113	=	Newton-meters
ENERGY OR WORK				
Btu	×	778.2	=	Foot-pounds
Btu	×	252.0	=	Gram-calories
POWER				
Btu per hour	×	.293	=	Watts
Horsepower	×	33000.0	=	Foot-pounds per minute
Horsepower	×	550.0	=	Foot-pounds per second
Horsepower	×	746.0	=	Watts
Kilowatts	×	1.341	=	Horsepower
PLANE ANGLE				
Degrees	×	.0175	=	Radians
Minutes	×	.01667	=	Degrees
Minutes	×	2.9×10^{-4}	=	Radians
Quadrants	×	90.0	=	Degrees
Quadrants	×	1.5708	=	Radians
Radians	×	57.3	=	Degrees

Pounds are U.S. avoirdupois.
Gallons and quarts are U.S.

102A02.TIF

TEMPERATURE CONVERSION TABLE

Locate the known temperature in °C/°F column. Read the converted temperature in °C or °F column.

°C	°C/°F	°F	°C	°C/°F	°F	°C	°C/°F	°F
−45.4	−50	−58	15.5	60	140	76.5	170	338
−42.7	−45	−49	18.3	65	149	79.3	175	347
−40	−40	−40	21.1	70	158	82.1	180	356
−37.2	−35	−31	23.9	75	167	85	185	365
−34.4	−30	−22	26.6	80	176	87.6	190	374
−32.2	−25	−13	29.4	85	185	90.4	195	383
−29.4	−20	−4	32.2	90	194	93.2	200	392
−26.6	−15	5	35	95	203	96	205	401
−23.8	−10	14	37.8	100	212	98.8	210	410
−20.5	−5	23	40.5	105	221	101.6	215	419
−17.8	0	32	43.4	110	230	104.4	220	428
−15	5	41	46.1	115	239	107.2	225	437
−12.2	10	50	48.9	120	248	110	230	446
−9.4	15	59	51.6	125	257	112.8	235	455
−6.7	20	68	54.4	130	266	115.6	240	464
−3.9	25	77	57.1	135	275	118.2	245	473
−1.1	30	86	60	140	284	120.9	250	482
1.7	35	95	62.7	145	293	123.7	255	491
4.4	40	104	65.5	150	302	126.5	260	500
7.2	45	113	68.3	155	311	129.3	265	509
10	50	122	71	160	320	132.2	270	518
12.8	55	131	73.8	165	329	136	275	527

$$°F = (9/5 \times °C) + 32$$
$$°C = 5/9 \, (°F − 32)$$

102A03.TIF

The NCCER makes every effort to keep these textbooks up-to-date and free of technical errors. We appreciate your help in this process. If you have an idea for improving this textbook, or if you find an error, a typographical mistake, or an inaccuracy in NCCER's *Contren®* textbooks, please write us, using this form or a photocopy. Be sure to include the exact module number, page number, a detailed description, and the correction, if applicable. Your input will be brought to the attention of the Technical Review Committee. Thank you for your assistance.

Instructors – If you found that additional materials were necessary in order to teach this module effectively, please let us know so that we may include them in the Equipment/Materials list in the Annotated Instructor's Guide.

Write: Product Development
National Center for Construction Education and Research
P.O. Box 141104, Gainesville, FL 32614-1104

Fax: 352-334-0932

E-mail: curriculum@nccer.org

Craft _____ Module Name _____

Copyright Date _____ Module Number _____ Page Number(s) _____

Description _____

_____

_____

_____

(Optional) Correction _____

_____

_____

(Optional) Your Name and Address _____

_____

_____

Survey Equipment Use and Care One

COURSE MAP

This course map shows all of the modules in the first level of the Site Layout curriculum. The suggested training order begins at the bottom and proceeds up. Skill levels increase as you advance on the course map. The local Training Program Sponsor may adjust the training order.

SITE LAYOUT LEVEL ONE

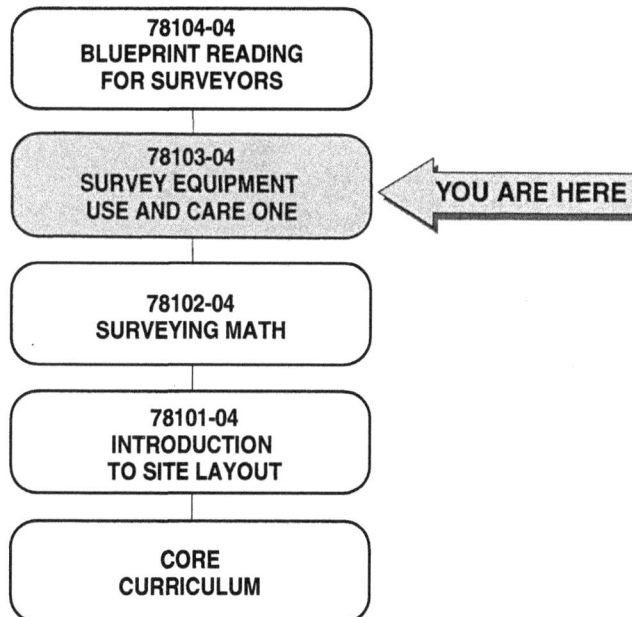

78104-04
BLUEPRINT READING
FOR SURVEYORS

78103-04
SURVEY EQUIPMENT
USE AND CARE ONE   ◁ YOU ARE HERE

78102-04
SURVEYING MATH

78101-04
INTRODUCTION
TO SITE LAYOUT

CORE
CURRICULUM

103CMAP.EPS

MODULE 78103-04 CONTENTS

Figures

Survey Equipment
Use and Care One

Objectives

When you have completed this module, you will be able to do the following:

1. Identify, safely use, and properly maintain the tools and instruments commonly used for site layout tasks.
2. Use a builder's level, transit, or theodolite and differential leveling procedures to determine site and building elevations.
3. Use accepted practices to record site layout data and information in field notes.
4. Check and/or establish 90° angles using the 3-4-5 rule.
5. Turn a 90° angle and double an angle.

Prerequisites

Before you begin this module, it is recommended that you successfully complete the following: Core Curriculum; Site Layout Level One, Modules 78101-04 and 78102-04.

Required Trainee Materials

1. Pencil and paper
2. Appropriate personal protective equipment

1.0.0 ◆ INTRODUCTION

This is the first of two modules that cover the use and care of the tools and instruments commonly used to perform site survey work. This module introduces the equipment used to perform distance measurements, **differential leveling**, and basic horizontal and vertical angular measurements. It also covers the guidelines and recommendations for recording surveying measurement information in **field notes**. Background information related to the equipment and site layout tasks covered in this module is also given.

2.0.0 ◆ SITE LAYOUT INSTRUMENTS AND EQUIPMENT

A wide variety of optical and electronic instruments are used to perform distance measurements, leveling, and angular measurement site layout tasks. The particular instrument used depends on the application or task and on the accuracy of measurement required. The following are commonly used site layout instruments:

- Builder's levels (covered earlier)
- Automatic levels
- Transits
- Optical theodolites
- Electronic transits and theodolites
- Total stations
- Electronic field books

2.1.0 Automatic Leveling Instruments

Automatic levels (*Figure 1*) are used to perform the same measurements and operations as described for the builder's level. These instruments have a built-in compensator mechanism that works to automatically maintain a true level line of sight. Compensator instruments still have to be leveled within the range of the compensator by three screws located on the base. An automatic leveling instrument must be kept upright and should never be carried over your shoulder. It contains a prism that can be damaged if the level is carelessly handled.

Figure 1 ♦ Automatic level.

2.2.0 Transits

The transit (*Figure 2*), commonly called a transit level, has been used for decades as the basic site layout instrument. Many transits of modern design are available today, but many older designs are still being used in the field.

Transits basically consist of three assemblies: the **alidade** assembly, the horizontal circle assembly, and the leveling head (*Figure 3*).

The alidade assembly, located at the top of the transit, includes the telescope, telescope level vial, vertical circle or arc, and vertical **vernier**. Also included are the plate leveling vials, telescope clamp and motion screw, and the vernier(s) for reading the horizontal circle. Depending on the model, transit telescopes can have powers ranging from 18 power (18X) to 32 power (32X). The power of a telescope determines how much closer an object will appear when viewed through the telescope.

The horizontal circle assembly located under the alidade contains the scale with which horizontal angles are measured, along with an upper motion clamp and related tangent screw. Some circle scales are marked from 0° to 360° and are numbered every 10° clockwise or counterclockwise. On others, the circle scale is divided into four quadrants, each reading from 0° to 90°. The leveling head located on the bottom of the transit under the circle assembly contains the leveling screws, lower motion clamp and related tangent screw, footplate, and baseplate. The baseplate provides for the attachment of the transit to a tripod. The baseplate at the base of the leveling head allows some movement of the transit proper in relation to the footplate when all of the leveling screws are loose. This aids in the positioning of the instrument directly over a point as the instrument is being set up.

The horizontal motion of the transit is controlled by upper and lower motion mechanisms, each consisting of a clamp and tangent screw. When tightened, the upper motion clamp joins or locks the alidade to the horizontal circle assembly, and the lower motion clamp locks or joins the horizontal circle to the leveling head. When a motion clamp has been tightened, the related tangent screw can be used to finely adjust the telescope position. Motion clamps should be finger-tightened only. Overtightening the clamps can strip the threads. The functions of the upper and lower motion clamps and tangent screws are as follows:

- With both the upper and lower motion clamps tightened, the entire instrument is immobile in relation to the leveling head and tripod.
- With the upper motion clamp tightened and the lower motion clamp loosened, the alidade and horizontal circle assemblies cannot move in relation to each other, but the joined assemblies and attached telescope can be rotated through 360° in the horizontal plane.
- With the lower motion clamp tightened and the upper motion clamp loosened, the same 360° sweep is allowed, but this time the vernier(s) on the alidade will pass by graduations on the horizontal circle. This allows measurement of horizontal angles between any given pointing of the telescope.
- When the upper or lower motion clamp is hand-tightened, the related tangent screw provides for a more precise horizontal setting of the transit.

Figure 2 ♦ Transit.

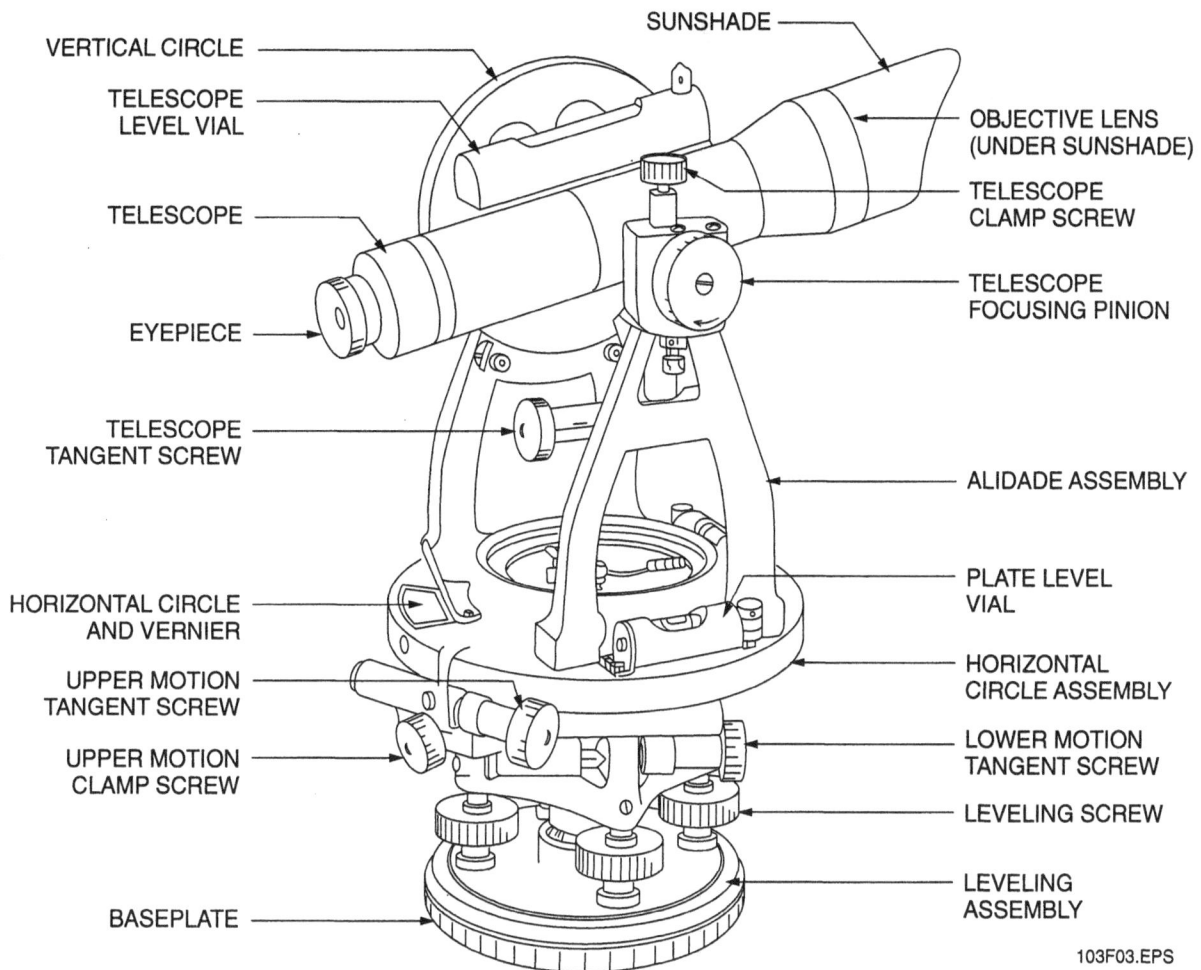

VERTICAL CIRCLE

TELESCOPE LEVEL VIAL

TELESCOPE

EYEPIECE

TELESCOPE TANGENT SCREW

HORIZONTAL CIRCLE AND VERNIER

UPPER MOTION TANGENT SCREW

UPPER MOTION CLAMP SCREW

BASEPLATE

SUNSHADE

OBJECTIVE LENS (UNDER SUNSHADE)

TELESCOPE CLAMP SCREW

TELESCOPE FOCUSING PINION

ALIDADE ASSEMBLY

PLATE LEVEL VIAL

HORIZONTAL CIRCLE ASSEMBLY

LOWER MOTION TANGENT SCREW

LEVELING SCREW

LEVELING ASSEMBLY

103F03.EPS

Figure 3 ◆ Transit assemblies and parts.

Unlike a builder's or automatic level, the transit level can be used for vertical as well as horizontal angles. The vertical motion of the transit controls the **vertical angle** of the telescope in the same manner as described for a horizontal motion.

2.3.0 Optical Theodolites

Optical theodolites (*Figure 4*) are used to perform the same site layout tasks as transits. Generally, theodolites are more precise instruments than transits and, as such, provide for greater accuracy in angular measurements. Typically, theodolites also have more automatic features than transits.

A theodolite is normally lighter and smaller than a transit. The telescope is shorter and does not have a spirit bubble level. Motion clamping systems on theodolites can differ greatly. Some theodolites operate in the same way as a transit. Others have only one motion clamp and tangent used for dual purposes. These instruments have a lever or switch that transfers clamp operation from upper to lower motion.

103F04.EPS

Figure 4 ◆ Optical theodolite.

A theodolite typically has a three-screw leveling head. The tripod upon which a theodolite is mounted is somewhat different from a tripod used with a transit. A tripod used with a theodolite has a flat base through which a bolt is threaded up into the three-screw leveling base (tribach), thus securing the instrument to the tripod.

In a theodolite, the graduated horizontal and vertical circles are fully enclosed within the instrument. These are viewed using a built-in optical system through which the user may read both horizontal and vertical angles through an eyepiece located near the telescope. They normally have an **optical plummet** that allows the operator to optically line up the instrument over a reference point, such as a hub, by looking through an eyepiece and aligning the crosshairs over the reference point. This enables the quick and accurate setup of the instrument over a point.

2.4.0 Electronic Transits and Theodolites

Electronic (digital) transits and theodolites (*Figure 5*) are widely used because they are dependable and easy to use. Unlike the optical instruments, there is no need to use microscopes, verniers, or micrometers to read angles. In these instruments, horizontal and vertical angle readings are monitored by self-checking microprocessors, electronically read, and then displayed digitally on the instrument's liquid crystal display (LCD). An advantage of using electronic transits/theodolites is that they save time and reduce errors when reading angles.

2.5.0 Total Station

A total station combines the functions of a theodolite, an electronic distance measurement instrument (EDMI), and an internal computer (data collectors) into a single instrument that can be used to make both distance and angular measurements (*Figure 6*). Total stations are commonly used to measure horizontal and vertical angles, measure **slope** distances, compute the horizontal and vertical components of these distances, and determine the coordinates of the observed points. Many designs and models of total stations are available.

The total station is microprocessor controlled and typically has a large built-in memory capability used to store data for thousands of layout points and/or data records. Under software control, two-way data flow between the total station and a remote location is possible via a standard

103F05.EPS

Figure 5 ◆ Electronic theodolite.

(RS-232) communications interface. This capability enables the data collected and stored in the instrument to be recalled and downloaded to a local or remote printer for hard copy or to a computer for calculations. It also enables data that has been processed at a remote location to be sent back to the instrument for field use. Because of the wide variety of total stations available, each one should be set up, aligned, and operated in accordance with the manufacturer's operating instructions. This procedure normally involves entering and/or downloading several bits of data into the instrument before or during the measurement. The use and operation of a total station is covered in more detail later in your training.

2.5.1 Types

Total stations are divided into two groups: manual and automatic. Total stations whose angle dials (circles) must be read optically are called manual instruments. Total stations with electronic circuits to read and display the distances and angles are called automatic instruments.

Total stations are further specialized by whether or not their distance-measuring components can be removed from the main instrument. When the distance-measuring device(s) can be removed from a total station, it is called a modular total station. If the distance-measuring devices cannot be removed from a total station, it is called a self-contained total station.

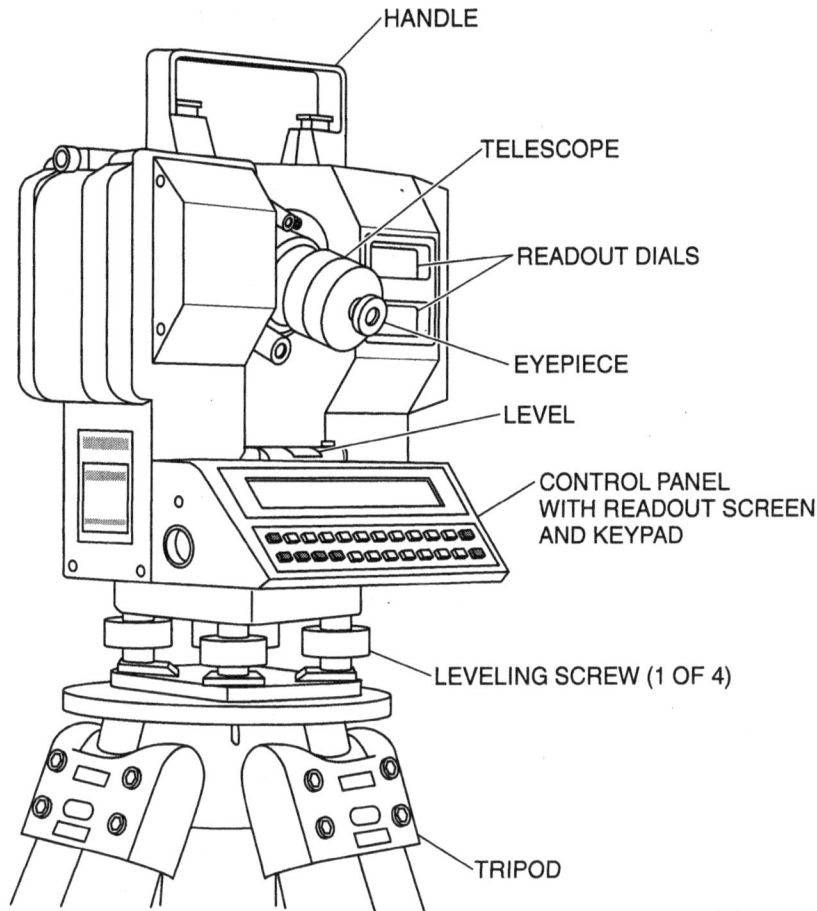

HANDLE
TELESCOPE
READOUT DIALS
EYEPIECE
LEVEL
CONTROL PANEL
WITH READOUT SCREEN
AND KEYPAD
LEVELING SCREW (1 OF 4)
TRIPOD

103F06.EPS

Figure 6 ◆ Total station.

2.5.2 Functions

The functions of a total station are to measure the line-of-sight (observed) distance from the instrument to a known point; to calculate the elevation of a point above or below the 0° reference line; and to calculate, in the horizontal plane and from a known reference point, the location of a point from the instrument.

The measured observed distance from the instrument to the point forms the hypotenuse of a right triangle associated with elevation views. If the telescope of the total station is angled above or below the level line at 0° to see the object being measured, the angular repositioning away from the 0° line creates the angle Θ for an elevation right triangle. Circuits within the total station use the observed distance and elevation angle to compute the elevation distance (opposite side) of the object above or below the level line. At the same time, the total station circuits are computing the ground range (adjacent side) of the point being measured.

If the total station has been aligned to a known reference point such as magnetic north, then its movement in the horizontal plane will create an

azimuth angle from the reference point. If the ground range from the instrument to the object being measured is already known, that ground range can be used as a hypotenuse, with the azimuth angle, to determine the east/west and north/south distances of the object from the instrument. Some total stations are able to compute the east/west and north/south distances of the object from the instrument; others do not have this capability.

The result is that total stations can do all the mathematical work for the operator. Operators who already understand basic surveying terms and math, as well the basics of transit operations, should be able to quickly learn how to operate any total station.

2.5.3 Controls

For the most part, the basic leveling and angle-positioning controls of a total station are very similar to those found on a transit. Total stations mounted on top of a tripod must still be initially leveled before the self-leveling devices inside the instrument take over.

The operator still must aim the telescope (optics) of the total station toward the target, which means that the body of the total station must be rotated horizontally until the target can be seen through the eyepiece of the telescope. As the front end of the total station's telescope is moved up or down to center the target in the eyepiece, the line-of-sight reference line of the telescope is moving in relation to the level line.

After the optics of the total station are aligned onto a target, control keys on the total station's keypad are used to make the total station read the observed distance to the target. Depending on the kind of total station being used, the elevation and ground range distances may also be computed and displayed by the station. Some of the newer total stations can also store the recorded data so that it can be recovered later or possibly be downloaded into a computer. Regardless of the capabilities of a total station, operators should still record the readings taken by a total station and then do their own math calculations to verify the final readings of the total station.

The controls and capabilities differ with each make and model of total station encountered. Operators will need to study the individual operator's manual for each instrument to determine its specific controls and capabilities.

2.5.4 Uses

Levels are good for determining the elevation of a given point in reference to the elevation of the level, but simple levels cannot provide specific angular information in either the vertical or horizontal planes. Theodolites or transits are modified levels that can be moved in specific directions from given reference points and then held in place until angular information, in either the vertical or horizontal planes, can be documented. The problem with both levels and transits is that they still have to be used with some kind of distance-measuring device such as a measuring tape. With both levels and transits, the math calculations have to be completed manually by the operators operating the instruments.

With simple EDMIs, line-of-sight distances can be measured, but that is about the extent of what an EDMI can do. If an EDMI is used with a transit, the opposite and adjacent sides of right triangles can be computed, even if they have to be determined manually.

With the invention of total stations, EDMIs and theodolites or transits were combined into one instrument that can electronically do all the math calculations for the operator and then provide the results on an analog or digital readout that is part of the total station instrument.

2.5.5 Care and Maintenance

The total stations on the market today, like most other kinds of surveying instruments, are built to withstand a fair amount of abuse before becoming inaccurate or nonoperational, but users of total stations must treat the instruments as well as possible to prevent undue damage.

Electronic instruments such as total stations need to be kept clean and protected from the elements such as excessive sunlight, rain, or snow. They also need to be calibrated when possible. Electrical circuits tend to change slightly with age and use. Having a base line of data to compare the instrument against ensures that it does not get so far off calibration that it could affect the daily readings.

2.6.0 Electronic Field Books

Electronic field books (*Figure 7*) are software/microprocessor-controlled devices normally used in conjunction with total stations and other electronic surveying instruments. They can be hand-held or attached to a tripod with a special bracket. Electronic field books are made with memory storage capacities typically ranging up to 4 megabytes (MB). An electronic field book is used to collect, store, and display the measured data obtained by the instrument to which it is attached. When desired, measurement data and descriptive notes can be entered manually via the keyboard. Data stored in the field book can be recalled and downloaded to a local or remote printer for hard copy or to a computer for calculations. Similarly, data that has been calculated, analyzed, and edited at a remote location can be sent back to the electronic field book for field use.

2.7.0 Care and Handling of Instruments

Instruments should always be maintained and handled in accordance with their manufacturer's instructions. The following are general guidelines for the proper use, care, and handling of all instruments:

- Only use an instrument if you know how to operate it.

- Keep the instrument in its carrying case when it is not in use.

- Handle the instrument by its base when removing it from the case or attaching it to the tripod.

- Never force any parts of the instrument. All moving parts should turn freely and easily by hand.

103F07.EPS

Figure 7 ◆ Electronic field book.

- Keep the instrument clean and free of dust and dirt. Clean the objective and eyepiece lenses using a soft brush or lens tissue. Rubbing with a cloth may scratch the lens coating and impair the view. Clean the instrument with a soft, non-abrasive cloth and mild detergent.
- Do not disassemble the instrument.
- Keep the equipment as dry as possible. If it gets wet, dry it before returning it to its case.
- When moving a tripod-mounted instrument, handle it with care. Carry it only in an upright position. Never carry it over your shoulder or in a horizontal position.
- Periodically have the instrument cleaned, lubricated, checked, and adjusted by a qualified repair facility or by the manufacturer, as specified in the manufacturer's instructions.
- Total stations must not be sighted on the sun unless filters are used. Sighting on the sun without filtering can damage the internal components of these instruments.
- For precision work, total stations should be shaded from direct sunlight and protected against very high temperatures and sudden changes in temperature.

3.0.0 ◆ LASER INSTRUMENTS

Laser instruments are widely used today for making site layout and other construction measurements. The use of modern construction laser

instruments provides many benefits, including the following:

- Ease of use
- Greater productivity
- Increased accuracy with a reduction in measurement errors
- Automatic instrument leveling and out-of-level beam cutoff

The laser devices used in modern construction instruments are typically laser diodes or helium-neon lasers. Laser diodes are semiconductor electron devices similar to the light-emitting diodes (LEDs) commonly used in digital instrument displays. The difference is that the laser diode converts an electrical input directly into coherent optical output power, whereas the LED emits incoherent light. Laser light is coherent light because it consists of a single wavelength. Incoherent light, such as that produced by a light bulb, consists of many different wavelengths. Because laser light is coherent, the beam can travel great distances without separating from a tight beam pattern.

Depending on the type of laser diode being used, the laser beam produced may be invisible to the human eye. Visible laser beams have wavelengths in the visible light portion of the radiant energy spectrum. Invisible laser beams have wavelengths in the infrared portion of the radiant energy spectrum. Because these beams are invisible, they must be detected using electronic laser detectors. Electronic laser detectors are described in detail later in this section.

A helium-neon laser is basically an electron tube filled with a combination of helium and neon gas. It produces a very intense, narrow beam of coherent red light that retains its shape over long distances. The beam can be manipulated, controlled, and detected quite easily. By expanding or focusing the beam with simple optical elements, the desired spot size can be achieved at almost any distance.

3.1.0 Construction Laser Instruments

Modern laser instruments used for construction layout work are typically battery-operated units that generate either a fixed laser beam, a rotating beam, or both. Fixed laser beam instruments can generate either single or multiple beams. One example of a fixed beam device is a simple carpenter's level with a built-in visible beam laser. It can be used as an ordinary carpenter's level or switched on to emit a laser beam, thereby extending the level's reference line typically between 200' and 300'. It can be mounted on a tripod or placed on a level surface. Another example of a single beam device is a handheld distance meter (*Figure 8*) used to measure distances up to about 100' without a target, and even farther distances with a reflective target. The auto-leveling laser alignment tool is a good example of a multiple-use laser instrument. It can be used to provide simultaneous plumb, level, and square reference points.

Rotating beam laser level instruments (*Figure 9*) are widely used for establishing grades and leveling over long distances. Both visible beam and invisible beam instruments are made in manual, self-leveling, and self-plumbing models. Most self-leveling laser level instruments have a capability that automatically turns off the laser beam whenever the instrument is not leveled. This feature is especially helpful if the instrument is accidentally bumped out of position. Should this happen, the unit automatically stops rotation of the laser beam and shuts off the beam. The beam remains shut off until the instrument is reset by the operator and leveled again. The instrument head can be rotated rapidly or slowly, or it can be stopped and used as a single beam device. One advantage of using a rotating laser instrument is that it allows one person, instead of two, to perform many site layout operations. After the instrument has been set up and leveled, the laser head can be adjusted to rotate at various speeds, typically ranging between 0 and 600 revolutions per minute (rpm), depending on the model. This creates a level, horizontal beam of light that sweeps over a 360° area so that it can be detected by multiple electronic laser detectors placed at different locations around the job site.

LASER DISTANCE METER

103F09.EPS

Figure 9 ◆ Rotating beam laser instrument.

The procedures for determining and establishing elevations with laser levels are similar to those used with the optical leveling instruments. To determine elevations, the base of the rod is held to the elevation to be determined. The laser detector is moved up or down on the rod until its display indicates it is centered on the beam. The reading on the rod is then recorded. An example of this is shown in *Figure 10*, where a rotating laser is being used to check the elevations at the four corners of a simple building foundation. As shown, a rod with a detector attached is placed at each of the four corners where the elevation readings are desired. The rotating laser beam will hit the detector at each location when the detector is in line with the beam. Similarly, to establish elevations, an electronic detector is attached to a grade rod. The grade rod is moved up or down until the beam indicates that the base of the rod is

LASER PLUMB, LEVEL, AND SQUARE
ALIGNMENT TOOL

103F08.EPS

Figure 8 ◆ Fixed beam laser instruments.

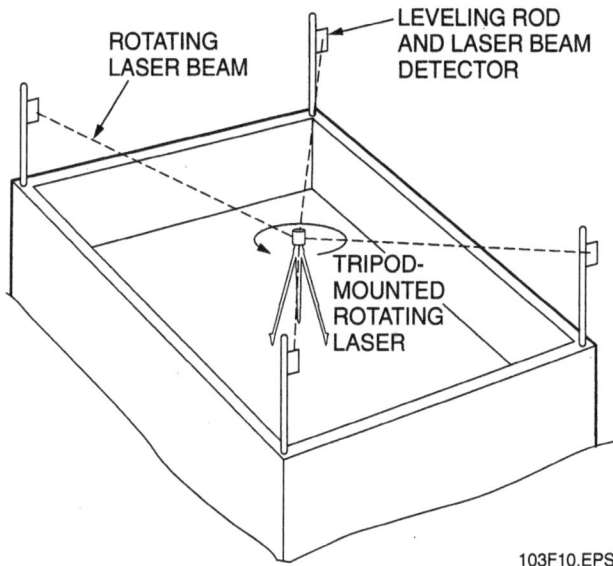

Figure 10 ◆ Rotating laser head used to sweep multiple sensors.

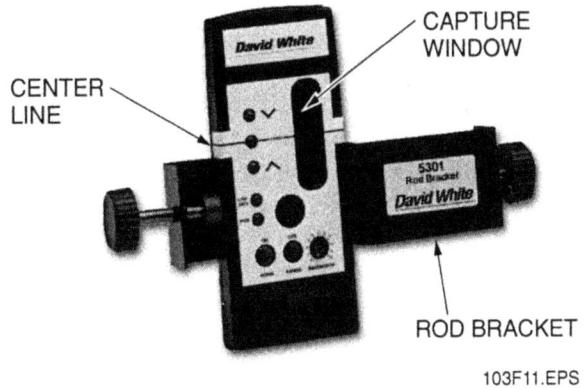

Figure 11 ◆ Electronic laser beam detector.

at grade. Most laser levels are also designed to work when positioned sideways using special mounting brackets or feet. This allows them to be used to perform vertical applications such as plumbing and aligning partitions, walls, formwork, and similar objects.

Another type of laser, called a pipe laser, is a special type of rugged laser instrument typically used by underground contractors when laying sewer or drain pipe. Unlike rotating lasers that create a plane of reference, pipe lasers provide a single beam (static) long distance laser beam that is used as a point of reference to either align the pipe or to determine pipe dropoff.

3.2.0 Electronic Beam Detectors

An electronic laser beam detector or sensor (*Figure 11*) is used in conjunction with laser instruments. It must be used to locate the laser beam line or plane when using an invisible beam (infrared) laser instrument such as an electronic level. In bright sunlight, it is also widely used to detect the beam of a visible light laser instrument. The laser beam detector is normally mounted on a leveling rod or similar support; however, handheld models are also available. During use, the detector is moved up or down on the rod to intersect the laser beam. A capture window on the beam detector receives the beam from the laser. The operation of detectors made by different manufacturers is basically the same. Typically, they have an LED visual display that indicates the position of the detector (down, on center, or up) in relation to the laser beam. Most also have an audible device that gives

an indication of beam position. For example, one manufacturer's detector uses a fast beeping audible tone to indicate that the detector should be moved down, a slow beeping tone meaning it should be moved up, and a steady continuous tone to indicate that the beam is on center. Some detectors also have bandwidth and sensitivity adjustments used to adjust the detector for various conditions including the degree of ambient light, laser wavelength, focus, power output, and bandwidth tightness required.

3.3.0 Use and Selection of Construction Lasers

General procedures for the setup and operation of laser levels to perform horizontal leveling and vertical plumb tasks differ with each manufacturer and model. You should set up and operate the instrument in accordance with the instructions given in the manufacturer's operator's manual for the specific instrument being used.

Using a laser level is fairly simple, but mistakes can still be made that will result in measurement errors. These include the following:

- Failure to follow good leveling practices
- Using instruments that are out of calibration
- Setting elevations incorrectly with the rod
- Performing laser measurements in the vicinity of other laser devices, which can cause unwanted stray signals to be received by the laser detector

When selecting a laser instrument for a particular application, the following factors should be taken into consideration:

- *Range* – The range of the instrument is normally determined by the size of the job site.
- *Power source* – Battery-operated lasers offer portability and do not require additional power unless recharging is needed.

- *Accuracy* – All lasers have an accuracy based on the design, range, and distance involved. For very precise work, a more accurate, self-leveling laser may be required, even if used for short-range applications.
- *Horizontal or vertical angles* – Some lasers are designed for use in horizontal applications only. If vertical plumbing, aligning, or measuring is involved, a laser with vertical capabilities is required.
- *Automatic leveling* – Automatic self-leveling lasers are easier and faster to set up and maintain accuracy on the job site. In order to prevent errors, these instruments normally have a built-in warning system that automatically cuts off transmission of the laser beam should the instrument become out of level for some reason.

3.4.0 Laser Instrument Safety

Under no circumstances should an unqualified worker operate a laser instrument. You must be properly trained as required by OSHA, and the training must be documented before you can operate a laser beam instrument. Government regulations also require that laser manufacturers provide warnings and cautions on their instruments and in their literature regarding the hazards associated with the use of laser instruments. Qualified personnel must adhere to all such manufacturer's warnings and cautions. They must also follow the manufacturer's operating instructions exactly when using a laser instrument.

> **WARNING!**
> Some laser units emit a very powerful and highly focused beam. Direct eye exposure to the laser beam can seriously injure the eyes or even cause blindness. This can also happen if the laser beam is reflected into the operator's eyes from a bright object such as a piece of shiny metal or a mirror. Eye damage occurs most often when the laser beam is stationary and a high-power laser is used.

If a rotating laser level is operated in the rotational mode, the beam will usually not harm the eyes because a lower power will be used. Most lasers used on construction sites are low-power lasers. The laser equipment will have a label on it that indicates the maximum power output, usually under five milliwatts, which is considered low power.

OSHA 29 CFR Part 1926 provides safety precautions that must be observed when a laser is in use. Manufacturers also provide various warnings and regulations to prevent hazardous situations. Some of these regulations are as follows:

- Prominent warning signs must be posted in the area where a laser is being used. Be alert to all such warning signs.
- Avoid going into areas marked off for laser operation unless it is necessary because of the job.
- Only those employees who have been properly trained are permitted to set up or operate a laser or make any adjustments to it.
- Qualified employees must carry an operator's card at all times when operating laser equipment.
- Always read and follow the manufacturer's operating instructions for the instrument being used.
- Avoid looking directly at the beam. Also avoid looking at any surfaces such as polished metal or a mirror, which can cause the beam to be reflected into the eyes.
- Special hazards are present when the laser is used in a fixed position, such as when it is used for tunnel or sewer work. Never look at the concentrated beam. Never look along the **axis** from the laser toward the point being sighted because the beam can be reflected from another surface.
- Wear approved safety goggles when the laser has a power output of 5 milliwatts or greater. The laser equipment must have a label indicating the maximum power output.
- Never point a laser beam at anyone. When possible, set the laser up so that it is above or below eye level.
- The laser beam should be turned off, shuttered, or capped when not in use.
- If you suspect you may have a vision problem caused by the laser, such as a persistent after-image, report for immediate medical attention.

3.5.0 Calibration and Care of Laser Instruments

The calibration of a laser instrument can be altered as a result of severe shock and vibration. Laser instruments should therefore be checked for proper calibration at regular intervals, particularly before starting on a new construction site. Calibration of the instrument should be performed by a qualified person and in accordance with the manufacturer's instructions. When an instrument is calibrated in the field, errors can be introduced because of temperature, humidity, and wind conditions. For this reason, it is recommended that the instrument be checked and calibrated in a controlled environment similar to that in which it will be used.

Although it is typically rugged enough for construction site work, a laser is a precision instrument requiring the same care as any high-quality leveling device. Observing the following maintenance guidelines will help ensure trouble-free laser operation:

- Always follow the manufacturer's recommended maintenance procedures as directed in the operator's manual for the instrument.
- To prevent moisture and dirt from settling inside the unit, always make sure the laser, accessories, and carrying case are clean and dry before storage. To clean the laser, electronic detectors, etc., wipe them off with a soft, non-abrasive cloth and clear, solvent-free water.
- When not in use, store the laser and its accessories in their original cases.
- Make sure the batteries in the unit and any spare batteries are fully charged in order to prevent work delays resulting from discharged batteries. If using rechargeable nickel cadmium batteries, also make sure that a battery charging unit is available to recharge the batteries when they become discharged.
- During periods of prolonged storage, it is advisable to set up and operate the laser for a minimum of eight hours per month.
- Do not attempt internal repairs to a laser instrument. Return it to a dealer, manufacturer, or other qualified organization for repair.

4.0.0 ◆ READING TRANSIT/ THEODOLITE SCALES AND VERNIERS

The values for the angles measured by transits, theodolites, and other instruments are indicated in several different ways, depending on the type of instrument, its make, and the model. Common ways used by instruments to indicate the angular values are vernier scales, optical scales, and digital displays.

4.1.0 Understanding Degrees, Minutes, and Seconds

Before discussing how to read angular values using the scales and verniers associated with the different types of instruments, it is important to review the following relationships relative to angular measurements. A circle is divided into 360 parts with each part called a degree; therefore, one degree = $\frac{1}{360}$ of a circle. The degree is the unit of angular measurement commonly used for surveying and site layout work. The degrees are further divided into minutes and seconds, where one degree is equal to 60 minutes (60') and one minute is equal to 60 seconds (60"). This arrangement, called the **sexagesimal system**, allows angles to be expressed in terms of degrees, minutes, and seconds. The sexagesimal system is used in the United States and most of the rest of the world to express angular measurements. However, some European countries express angular measurements using the centesimal system, in which a circle is divided into 400 parts called gons or grads. Also, in some military science work, angular measurements are based on a system that divides a circle into 6,400 parts called mils.

4.2.0 Reading Vernier Scales

Horizontal and vertical angle vernier scales used on non-electronic transits to indicate the values of angular measurements are the hardest to interpret and read accurately. Even though these instruments are being replaced by newer and easier to use instruments, many are still in use. For this reason, you need to understand how to read their vernier scales. A magnifying glass may be required to read these scales.

4.2.1 Reading Horizontal Angles

To read horizontal angles, the transit's horizontal circle and vernier scales are used. Depending on the transit being used, the horizontal circle scale can be graduated in different ways. Most circle scales are labeled from 0° to 360° and are graduated to the nearest 15, 20, or 30 minutes of arc. On some transits, the 360° circle scale is divided into four equal 90° quadrants, with each quadrant graduated from 0° to 90°. The transit's horizontal vernier scale is used to obtain more precise readings. Vernier scales are graduated in minutes and typically can be read to either 20 or 30 seconds of arc, depending on the design. *Figure 12* shows the relationship between degrees, minutes, and seconds represented by the graduations on a simplified transit vernier.

One widely used style of vernier scale is the double direct vernier (*Figure 13*). The graduated circle scale used with this vernier is marked with two sets of numbers, one increasing in a clockwise (CW) direction from 0° to 360° and the other increasing from 0° to 360° in a counterclockwise (CCW) direction. This allows the vernier to be read both in the CW and CCW directions, depending on whether the eye end of the telescope is turned to the left or right, respectively. On some transits, the horizontal circle may be read by two verniers (verniers A and B) which are mounted 180° apart on the transit alidade.

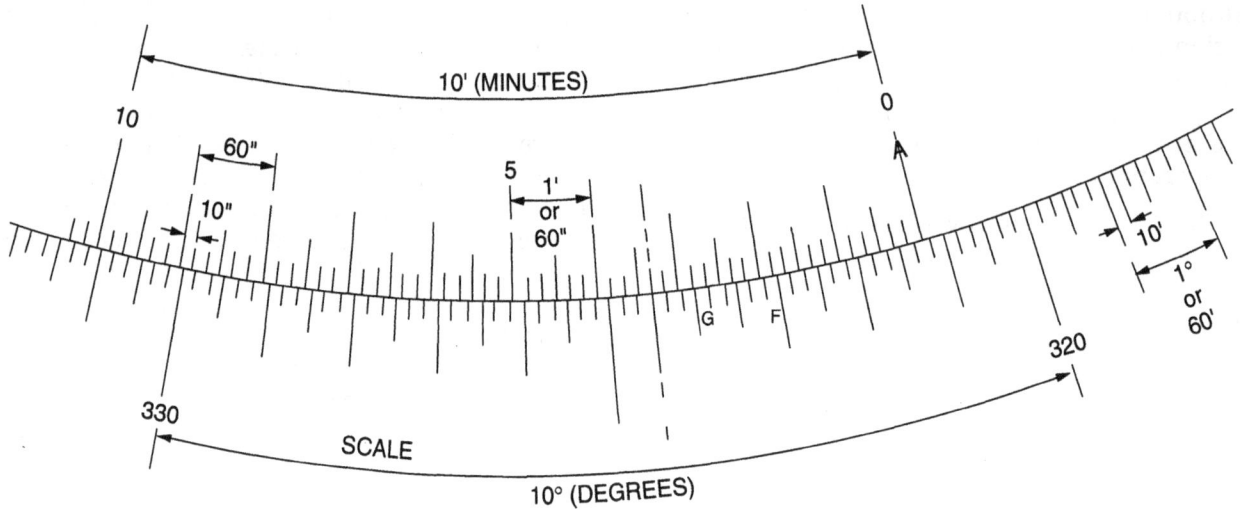

READING SHOWN IS 321°13'20"

103F12.EPS

Figure 12 ◆ Graduations on a typical transit vernier.

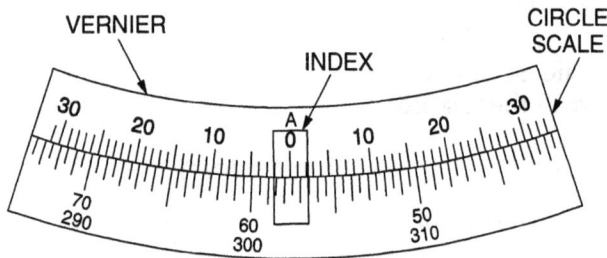

CLOCKWISE = 57°30'
COUNTERCLOCKWISE = 302°00'

(A) READING CIRCLE SCALE

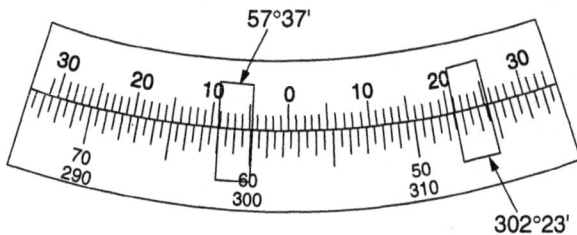

CLOCKWISE = 57°37'
COUNTERCLOCKWISE = 302°23'

(B) READING VERNIER SCALE

$$\begin{array}{r} 57°37' \\ 302°23' \\ \hline 359°60' = 360° \end{array}$$

(C) CHECK   103F13.EPS

Figure 13 ◆ Reading a double direct vernier.

The method for reading horizontal angles using a double direct vernier is described here. Reading other types of verniers is basically the same. To determine the degrees, minutes and seconds for an angular measurement, the readings from the main circle scales and vernier scales are added together. *Figure 13* shows an example of how to read a double direct vernier.

As shown in *Figure 13A* the index mark on the vernier shows that a CW angle of 57°30' has been passed on the circle scale when read in the CW direction. When read in the CCW direction, the vernier index shows that an angle of 302° has been passed.

To obtain a more precise measurement, the vernier degree scale reading is added to the circle degree reading. The vernier is read by finding a graduation on it that coincides (matches) with any division on the circle scale. On a double vernier there are two such matching lines, one for the CW angle and the other for the opposite CCW angle. Note that the graduations on a vernier are very fine and closely spaced together. For this reason, a magnifying glass may be needed in order to read the vernier accurately on older style transits.

The vernier shown in *Figure 13B* is graduated in 30 minutes capable of reading to one minute. As shown in *Figure 13B*, the vernier graduation that matches the circle graduation for a CW angle is 7'. Therefore, the CW angle being read is 57°37' (57°30' read on the circle scale +7' read on the vernier scale). The vernier graduation that matches the circle graduation for a CCW angle is 23'. Therefore, the CCW angle is 302°23' (302° + 23'). It should be pointed out that the sum of the CW and CCW angles should always equal 360°, otherwise an error has been made. To check the example, 57°37' + 302°23' = 359°60' = 360°. *Figure 14* shows two more examples of double direct vernier styles.

4.2.2 Reading Vertical Angles

To measure vertical angles, transits have either a vertical circle scale or a vertical arc scale. Both types of scales move with the tilting motion of the telescope. Like horizontal circle scales, vertical circle scales are graduated from 0° to 360°. Vertical arc scales are graduated from 0° to 45° in two directions. A vertical vernier scale that is used in conjunction with the vertical circle/arc scale is attached to the frame and does not move when the telescope is tilted. Vertical angles measured with a transit are referenced to the horizon. When the telescope spirit level is centered, the vernier index on the vertical circle/arc should read 0°00'. Following this, the telescope is tilted up or down

to the desired vertical measurement angle and the angle is read using the vertical circle/arc and vernier scales. The manner in which the scales are read is basically the same as described previously for reading horizontal angles. Some mistakes commonly made while reading verniers include the following:

- Reading the wrong direction from zero
- Not using a magnifying glass for those models that require one
- Misinterpreting the scale graduation values
- Omitting 10, 15, 20, or 30 minutes on the circle scale when the index is beyond these marks
- Failing to read directly on the line (**parallax**)

4.3.0 Reading Optical Scales and Digital Displays

When compared to instruments with vernier-type scales, instruments with optical scales or electronic instruments with a digital display are simpler to use. For an electronic transit or theodolite

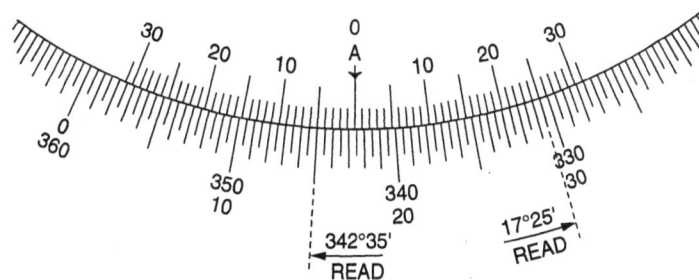

GRADUATED 30 MINUTES READING TO ONE MINUTE

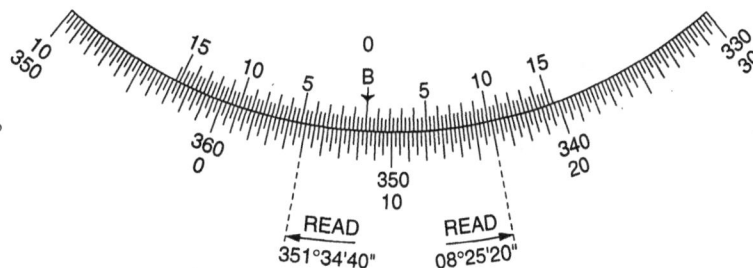

GRADUATED 15 MINUTES READING TO 20 SECONDS

103F14.EPS

Figure 14 ◆ Double direct vernier readings.

with a digital display, the value of the measured horizontal or vertical angle is read directly in the same way you would read the display on a pocket calculator. *Figure 15* shows an example of an electronic transit/theodolite digital display.

Optical theodolites have specially designed scales that are viewed by looking through an optical microscope on the instrument. Depending on the design of the theodolite, there are different kinds of scales: scale reading, repeating, or directional. These scales are relatively easy to use and should be read as described in the operator's manual for the instrument being used. A brief explanation and examples of three common types of optical readouts are given here.

4.3.1 Reading a Scale Reading Theodolite

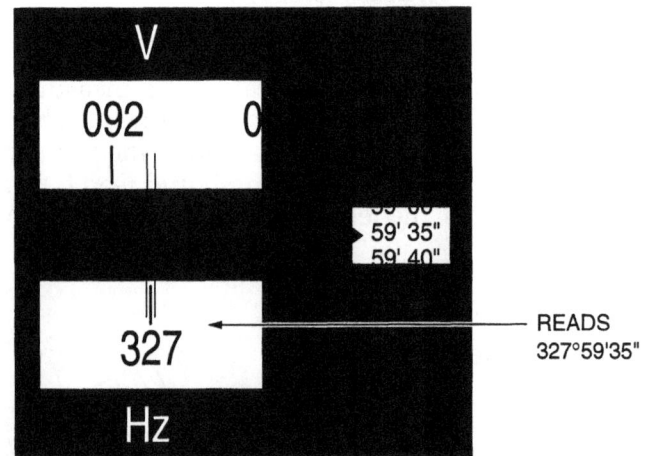

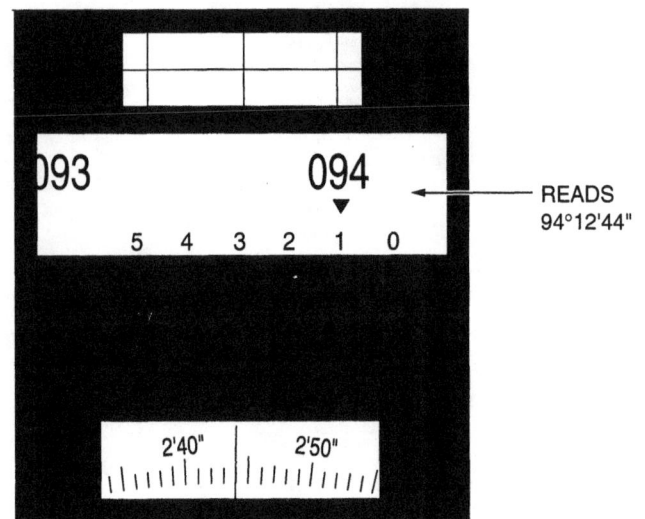

With a scale type of theodolite display, the vertical and horizontal degree circles are read directly with an optical microscope. As shown in *Figure 16A*, the scale is read where it is intersected by the degree reading from the circle.

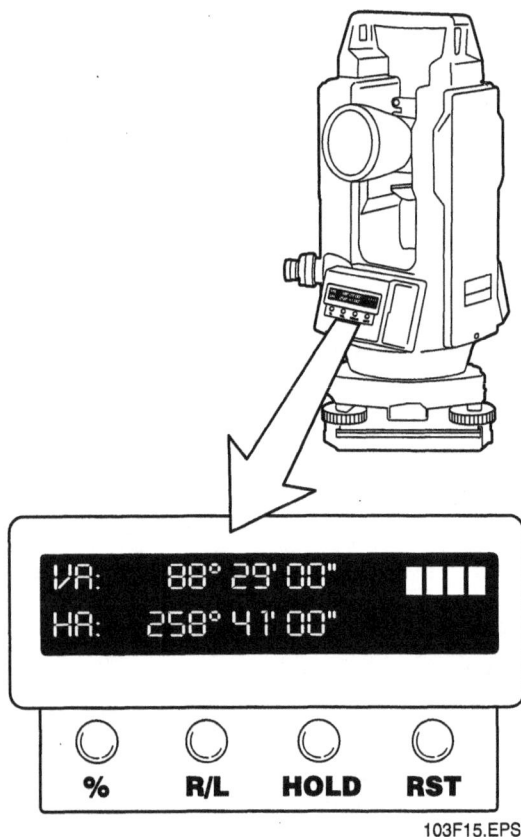

Figure 15 ◆ Electronic transit/theodolite display.

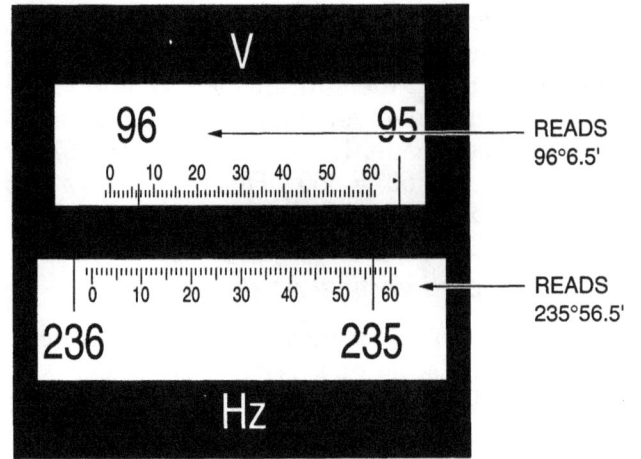

Figure 16 ◆ Common optical theodolite scales.

4.3.2 Reading a Repeating Theodolite

A repeating theodolite uses a glass degree circle and micrometer to read the turned angle. The micrometer is first used to align the degree index marks. Then the reading from the degree window and micrometer window are added together to obtain the angle, as shown in *Figure 16B*.

4.3.3 Reading a Directional Theodolite

The directional theodolite does not have a lower motion clamp and tangent screw. The horizontal circle remains fixed during a series of readings. The telescope is sighted on each of the measurement points and the directions are read. Horizontal angles are determined by calculating the differences between the directions. The reading of the angle is similar to that of the repeating theodolite. A micrometer is used to align the degree index marks. Then the reading from the degree window and micrometer window are added together to obtain the angle, as shown in *Figure 16C*.

5.0.0 ◆ INITIAL SETUP, ADJUSTMENT, AND CHECKOUT OF A TRANSIT/THEODOLITE

The initial setup of a transit and theodolite are basically the same. This section outlines the procedures for setting up a transit or theodolite directly over an established (fixed) point. This point will be the vertex for a subsequent horizontal angle measurement. Procedures are given both for setting up an older-style instrument over a point using a plumb bob and for setting up newer-style instruments over a point using a built-in optical plummet.

The initial setup of other site layout instruments such as builder's levels, automatic levels, and total stations is accomplished in basically the same way as described here for setting up a transit/theodolite.

    ◯
    ◯   **NOTE**
    ◯   Always keep one hand on the instrument while
 ((◯)) setting it up.

5.1.0 Setting Up Over a Point Using an Instrument with an Optical Plummet

As explained earlier, newer transits/theodolites are equipped with a device called an optical plummet. This device allows the instrument to be optically lined up over a reference point by looking through an eyepiece and aligning the crosshairs over the point. An optical plummet consists of a set of lenses and mirrors which enable the user of the instrument to look into a viewing port on the side of the instrument (*Figure 17*). The optics and mirrors are located in the lower part of the instrument so when the base of the instrument is perfectly level, the crosshairs of the optical plummet fall on a point exactly under the center of the instrument.

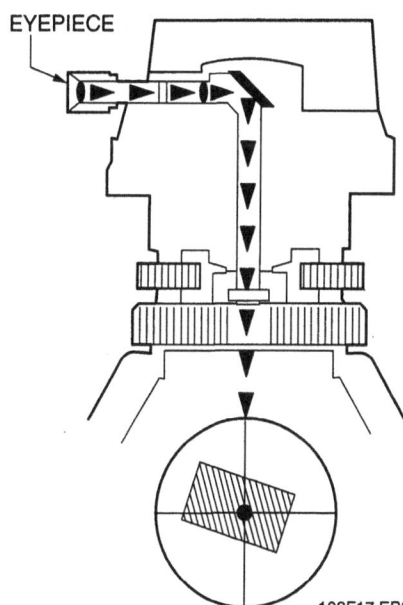

Figure 17 ◆ Optical plummet.

5.2.0 Checking Transit/Theodolite Calibration

Transits and theodolites should be field tested for correct calibration and adjustment if you are using an instrument for the very first time or if you suspect the instrument is out of adjustment. To help you understand the principles upon which calibration of an instrument is based, a brief overview of the geometry of angle-measuring instruments is given here. Some tests commonly performed to check the calibration of an instrument are also briefly described.

5.2.1 Geometry of Angle Measuring Instruments

Transits, theodolites, and all other instruments used to make angular measurements consist essentially of an optical line of sight, which is perpendicular (at a right angle or 90°) to and supported on a horizontal axis. As shown in *Figure 18*, the line of sight of the instrument is perpendicular to the horizontal axis, and the horizontal axis of the instrument is perpendicular to a vertical axis, about which it can rotate. The line of sight is perpendicular to the horizontal axis when the telescope level bubble is centered and when the vertical circle/arc is set at 90°/270°, or 0° for vernier transits. Spirit levels mounted on the base of the instrument alidade are used to make the vertical axis coincide with the direction of gravity. These geometric relationships must be maintained in the instrument; otherwise, the instrument will be out of calibration and any angles measured or laid out with it will be incorrect.

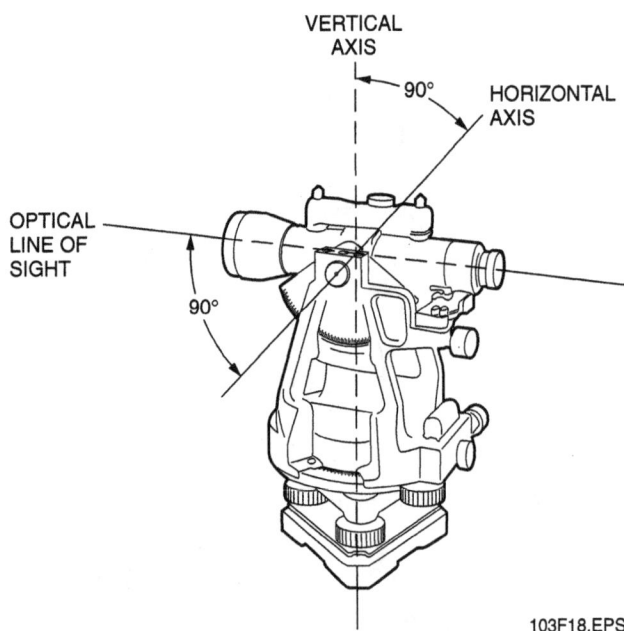

Figure 18 ◆ Geometry of an angle measurement instrument.

5.2.2 Plate Bubble Test

When in perfect adjustment, the plate bubbles on an instrument, once centered, should remain centered for all positions of the horizontal plate, unless the instrument settles or is otherwise disturbed. If either or both bubbles do not do this, adjustment is required. To determine whether the instrument requires adjustment, level the instrument carefully. Then, with one plate bubble centered over a pair of leveling screws, rotate the instrument through 180° in the horizontal plane.

The bubble should remain centered. If not, the instrument should be sent to a repair facility for adjustment.

5.2.3 Crosshair Tests

Crosshair tests are performed to make sure that the vertical and horizontal crosshairs are plumb and level, respectively. Both tests are easy to perform. The object of the horizontal crosshair test is to make sure that the instrument's horizontal crosshair is in a plane that is perpendicular to the vertical axis of the instrument. With a properly adjusted instrument, you should be able to place any part of the horizontal crosshair on the object or point being viewed with the telescope and still get an accurate reading. First, carefully level the instrument, then sight the horizontal crosshair on a distant target or other well-defined point (*Figure 19*). Once the crosshair is placed on the point, turn the instrument's horizontal tangent screw so that the instrument slowly rotates about its vertical axis. The horizontal crosshair should stay fixed on the point as the instrument is rotated. If any part of the crosshair moves above or below the reference point, the instrument needs adjustment and should be sent to a repair facility.

The object of the vertical crosshair test is to ensure that the instrument's vertical crosshair is in a plane that is perpendicular to the horizontal axis of the instrument. With a properly adjusted and leveled instrument, sight the vertical crosshair on a plumb bob string at rest. For a properly adjusted instrument, the vertical crosshair should coincide with the string. If any part of the crosshair does not completely coincide with the string (tilts to the left or right), the instrument needs adjustment and should be returned to a repair facility.

If either crosshair is out of adjustment, the instrument can still be used until the actual adjustment is made. This is done by using only that part of the crosshair which is nearest to the intersection point of both crosshairs.

5.2.4 Optical Plummet Check

The alignment of the optical plummet's axis relative to the vertical axis of a transit/theodolite should be checked periodically for accuracy. This is because an optical plummet can get out of adjustment, causing the instrument to be set over erroneous points.

The alignment of an optical plummet can be checked by placing its crosshairs over a reference point on the ground with the instrument at 0° (*Figure 20*). Following this, the instrument is rotated to 180°, and the position of the crosshairs is checked

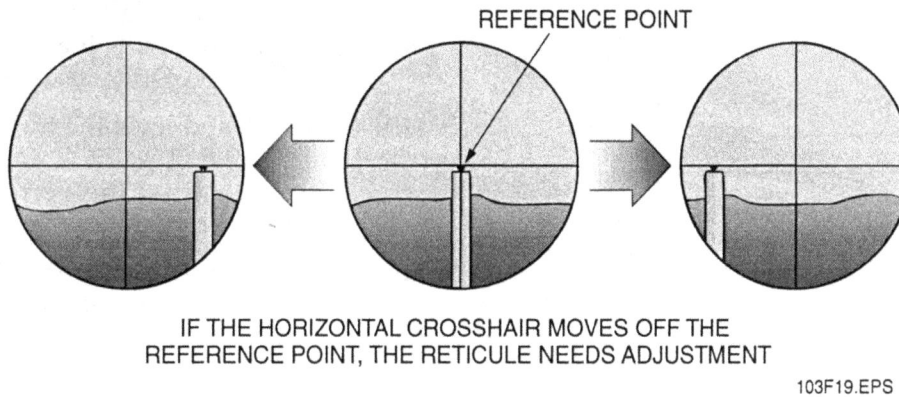

REFERENCE POINT

IF THE HORIZONTAL CROSSHAIR MOVES OFF THE
REFERENCE POINT, THE RETICULE NEEDS ADJUSTMENT

103F19.EPS

Figure 19 ◆ Horizontal crosshair test.

again. If the crosshairs are superimposed on the reference point at both positions, the plummet is aligned. If not, the plummet should be adjusted in accordance with the instrument manufacturer's instructions. Typically, this is done by turning the adjustment screws on the optical plummet so that its crosshairs are positioned over a point (Point A) midway between the original points sighted for the 0° and 180° positions. This procedure should be repeated as necessary so that the crosshairs remain superimposed on the reference point when the instrument is at the 0° and 180° positions.

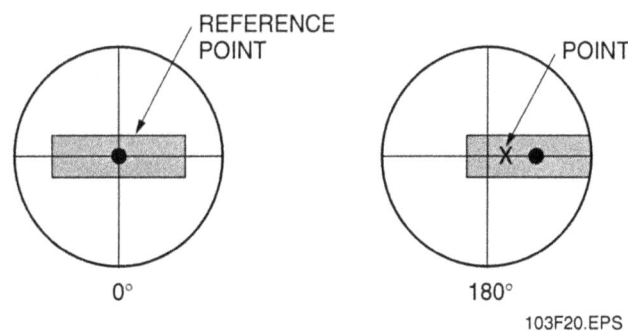

Figure 20 ◆ Optical plummet check.

5.2.5 Other Tests

Other calibration checks that can be performed on an instrument are listed below. These checks are normally performed by an authorized person in accordance with the instructions given in the instrument manufacturer's operator's manual, or described in several readily available reference books pertaining to surveying.

- *Line of sight test* – Determines if the instrument's line of sight is perpendicular to the horizontal axis.
- *Horizontal axis test* – Determines if the instrument's horizontal axis is perpendicular to its vertical axis.

- *Telescope bubble test* – Determines if the axis of the telescope's level is parallel to its line of sight.
- *Vertical circle vernier test* – Determines if the vertical circle reads zero when the instrument is properly leveled and the telescope bubble is centered.

6.0.0 ◆ BASIC HORIZONTAL AND VERTICAL ANGLE MEASUREMENTS

This section describes basic procedures for using instruments to establish and/or measure horizontal and vertical angles. For the purpose of explanation, the procedures given here describe the use of a standard engineer's transit equipped with two independent horizontal motions (upper and lower) and vernier scales. Depending on the type of instrument actually being used, the procedures will vary somewhat. However, with the exception of the specific instrument operating procedures, the general methods described here apply to measurements made with most instruments.

6.1.0 Turning 90° Angles

Turning 90° angles with a transit or other instrument is commonly performed when laying out the lines for a building foundation. For decades, building lines have been laid out using a transit and steel tape in a manner that traces the actual shape of the building foundation. *Figure 21* shows an example of this method of layout. The procedure for zeroing the transit and turning the 90° angles required to perform the layout task is given here. Line AB in the figure represents the lot line and point E one corner of the building. Point C is the point where the line forming one side of the building intersects the lot line.

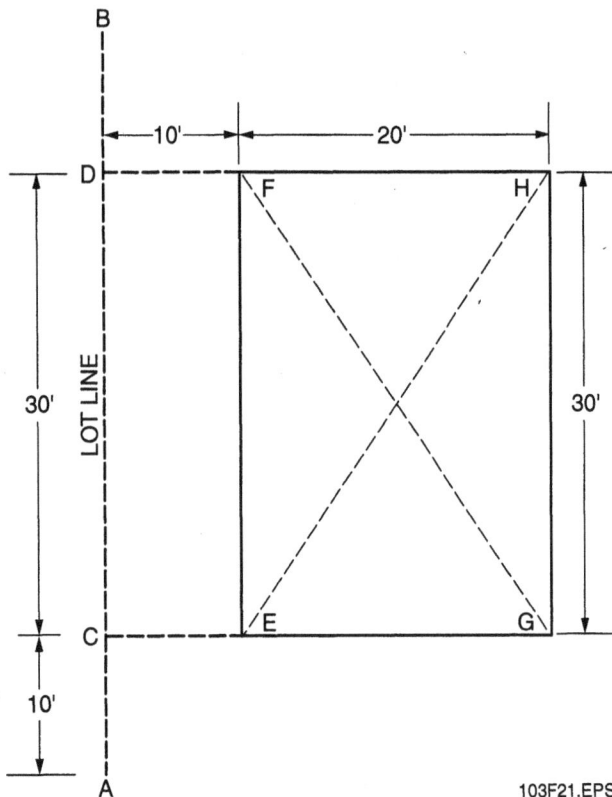

Figure 21 ◆ Site layout of building foundation lines.

Step 1 Set up and level the transit directly over a stake driven at point C. Sight the telescope so that the vertical crosshair is sighted on a rod or target that is on line with point B.

Step 2 Loosen both the upper and lower horizontal motion clamps, then turn the horizontal circle until the 0° graduation is nearly aligned with the vernier index. Tighten the upper motion clamp, then use the upper tangent screw to bring the horizontal circle 0° graduation into exact alignment with the vernier index. Once the vernier is set to zero, do not touch or loosen the upper motion clamp, since this can change the vernier setting.

Step 3 With the lower clamp loosened, rotate the telescope 90° clockwise (to the right) as indicated by the horizontal circle so that the vertical crosshair is sighted on line CEG. Tighten the lower motion clamp, then use the lower tangent screw to bring the horizontal circle 90° graduation into exact alignment with the vernier index.

Measure and lay out the distances from point C to points E and G along this line with a tape. Drive stakes at points E and G.

Step 4 Set up and level the transit directly over the stake at point G and sight the telescope so that the vertical crosshair is exactly on point E.

Step 5 Zero the transit horizontal circle and vernier as previously described in Step 2.

Step 6 With the lower clamp loosened, rotate the telescope 90° clockwise as indicated by the horizontal circle so that the vertical crosshair is sighted on line GH. Tighten the lower motion clamp, then use the lower tangent screw to bring the horizontal circle 90° graduation into exact alignment with the vernier index. Measure and lay out the distance from point G to point H along this line with a tape. Drive a stake at point H.

Step 7 Set up and level the transit directly over the stake at point H and sight the telescope so that the vertical crosshair is exactly on point G.

Step 8 Zero the transit horizontal circle and vernier as previously described in Step 2.

Step 9 With the lower clamp loosened, rotate the telescope 90° clockwise as indicated by the horizontal circle so that the vertical crosshair is sighted on line HF. Tighten the lower motion clamp, then use the lower tangent screw to bring the horizontal circle 90° graduation into exact alignment with the vernier index. Measure and lay out the distance from point H to point F along this line with a tape. Drive a stake at point F. This completes the layout. If the layout has been done correctly, the distances for lines CE and DF will be equal. Check the layout for square by measuring the length of diagonal lines EH and FG. These lines should be equal if the layout has been done correctly.

6.2.0 Measuring Horizontal Angles

To measure an unknown horizontal angle between two lines, proceed as follows. For the purpose of an example, assume that the instrument is located at point A (*Figure 22*), and it is desired to measure the angle between lines AB and AD.

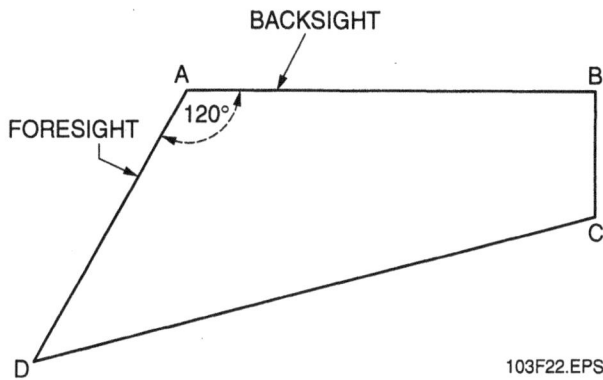

Figure 22 ◆ Measuring an angle.

Step 1   Set up and level the transit over the established reference point (point A).

Step 2   Loosen both the upper and lower horizontal motion clamps, then turn the horizontal circle until the 0° graduation is nearly aligned with the vernier index. Tighten the upper motion clamp, then use the upper tangent screw to bring the horizontal circle 0° graduation into exact alignment with the vernier index. Once the vernier is set to zero, do not touch or loosen the upper motion clamp or tangent screw because this can change the vernier setting.

Step 3   With the lower clamp loosened, rotate the telescope so that it is pointed at a target located at point B. Sight the telescope vertical crosshair so that it intersects the target, then tighten the lower motion clamp. Use the lower tangent screw to bring the vertical crosshair to an exact setting on the target. The line of sight for the telescope is now on one line or leg of the angle to be measured, and the instrument vernier scale is set to zero. This position, called the backsight, is the starting or reference point for the measurement of the angle.

Step 4   With the lower motion clamp still tightened, loosen the upper motion clamp and rotate the telescope so that it is pointed at a target located at point D. Sight the telescope vertical crosshair so that it is placed on the target; then tighten the upper motion clamp. Use the upper tangent screw to bring the vertical crosshair to an exact setting on the target.

This position, called the foresight, is the ending point or line for the measurement of the angle.

Since the horizontal circle is clamped by the lower motion clamp, the zero still points toward the initial point. Because the upper motion clamp was loosened while the telescope was being rotated to the foresight position, the vernier on the upper plate moved over the circle as the transit was turned, thus indicating the value for the angle turned.

Step 5   Read and record the value of the angle turned using the circle and vernier scales. Be sure to read both the circle scale and the vernier in the direction that the vernier passed over the scale. For our example, the angle measured is 120°00'.

> **NOTE**
>
> One method commonly used to check for mistakes and/or to increase precision is to use a procedure called measuring angles by reversing or doubling. This is done by first measuring an angle once, as described in Steps 1 through 5. To measure angles by reversing or doubling, proceed as described in Steps 6 through 9.

Step 6   Reverse or **plunge the telescope** (turn it upside down) by loosening the vertical motion clamp, then rotating it 180° around its horizontal axis.

Step 7   Loosen the lower motion clamp and rotate the telescope back to the initial backsight position (point B). Sight the telescope's vertical crosshair so that it intersects the target and tighten the lower motion clamp. Use the lower tangent screw to bring the vertical crosshair to an exact setting on the target. Note that the value of the first angle turned (120°) is still fixed on the vernier scale.

Step 8   Loosen the upper motion clamp, then rotate the telescope back to the foresight position (point D). Sight the telescope vertical crosshair so that it intersects the target, then tighten the upper motion clamp. Use the upper tangent screw to bring the vertical crosshair to an exact setting on the target.

Step 9   Read and record the value of the angle turned using the circle and vernier scales. The difference now is that the vernier indicates an angular value that is twice that of the first reading. For our example, the first angle reading measured 120°00', so the second reading should measure 240°00'. The values of the double angle and two times the value measured for the single angle should agree within tolerances. Also, this doubled angular value must be divided by two to obtain the average value for the angle measured.

6.2.1 Measuring Angles by Repetition

Measurement of angles by repetition is used when increased accuracy is required. It is used when it is desired to gain accuracy beyond the **least count** of the instrument being used. The least count is the finest reading that can be made directly on a vernier of a transit or micrometer of a theodolite.

Measurement of an angle by repetition is identical to the procedure described above for measurement by doubling except there are from four to eight repetitions made instead of only two. When recording the values measured for each angle, normal practice is to record only the first and last readings. Following this, the value for the accumulated (summed) angular measurements is divided by the number of repetitions to derive the average value for the angle. For example, assume that after six repetitions (three direct and three reversed) the summed angular value is 240°00'. The average value for the angle is then equal to 40°00' (240°00' ÷ 6). Note that it is often necessary to add 360°, or multiples of 360°, to the final instrument reading in order to account for the number of complete 360° revolutions the telescope has been turned horizontally while making the repeated measurements. For example, a 60°00' angle measured eight times causes the instrument to be turned through 480°00'; however, the instrument's scale would read only 120°00'. Therefore, to calculate the average value for the angle being measured, it is necessary to add 360° to 120°00' before dividing by 8 (480°00' ÷ 8 = 60°00').

6.2.2 Closing the Horizon

A technique called closing the horizon can be used to check the accuracy of angular measurements. Closing the horizon means that the unused angle is measured to complete the circle (*Figure 23*). When the horizon is closed and all angles at the station are added together, the sum should be

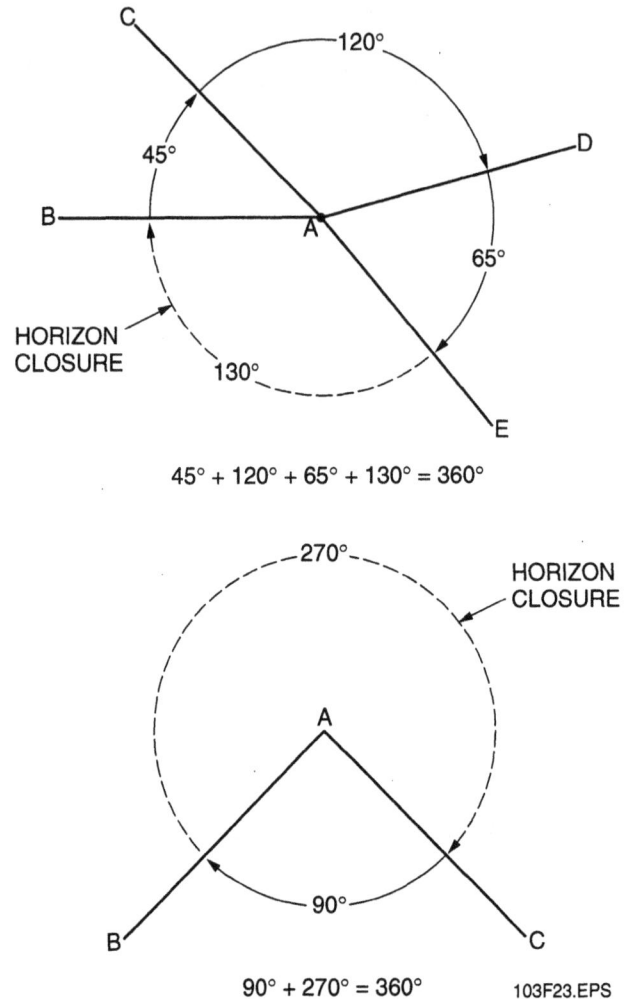

45° + 120° + 65° + 130° = 360°

90° + 270° = 360°   103F23.EPS

Figure 23 ◆ Simplified examples of horizon closures.

exactly 360°. Normally there will be some small error. Should the error be large (more than 30 seconds), a mistake has been made in the measurements, and they should be redone.

6.3.0 Measuring Vertical Angles

Vertical angles are measured in a similar way to horizontal angles. Vertical angles are measured with reference to the horizon when using a transit. A measurement from the horizon to a high point is a positive (+) vertical angle, and from the horizon to a low point is a negative (–) vertical angle. Obviously, it is important to record whether a vertical angle is positive or negative.

A transit must be carefully set up and leveled to measure a vertical angle. The telescope bubble should be centered and the vernier on the vertical vernier scale should read 0°00'. The procedure for measuring a vertical angle requires that the horizontal and vertical motion clamps be loosened.

The telescope is rotated and vertically positioned so that the horizontal crosshair rests approximately on the point to which the vertical angle is to be measured. With the vertical motion clamp or clamps tightened, the vertical tangent screw is adjusted to set the horizontal crosshair exactly on the point. Following this, the value of the angle is read from the vertical circle and vernier. Note that if the transit has a full vertical circle and the telescope can be plunged, a more accurate reading can be obtained by measuring the vertical angle twice, once with the telescope direct (upright position) and once reversed (inverted position), then averaging the two readings.

With some instrument models, the vertical angular reading with the telescope level is 90° instead of 0°. If this is the case, when measuring a positive vertical angle it is necessary to subtract the angle reading on the instrument from 90° to obtain the actual angle. For example, if the instrument angular reading is 72°30', the actual angle being measured is 17°30' (90°00' − 72°30').

When measuring a negative vertical angle it is necessary to subtract 90° from the instrument reading to obtain the actual angle. For example, if the instrument angular reading is 118°30', the actual angle being measured is 28°30' (118°30' − 90°00').

6.4.0 Common Mistakes Made When Making Angular Measurements

Some common mistakes made when making angular measurements include the following:

- Poor setup and leveling of instruments
- Misreading the instrument scale indications
- Transposing and/or recording the wrong angle values in field notes
- Sighting on the wrong targets, marks, or lines when measuring horizontal or vertical angles
- Using the wrong instrument tangent screw
- Failure to center the telescope bubble before measuring a vertical angle
- Failure to take into account the algebraic signs for the values of vertical angles measured with a transit

7.0.0 ◆ BASICS OF DIFFERENTIAL LEVELING

Differential leveling is the process used to determine or establish elevation grades such as those needed for setting slope stakes, grade stakes, footings, anchor bolts, slabs, and sidewalks. The procedural data relating to differential leveling given here emphasizes the use of conventional leveling instruments such as the builder's level or automatic level.

An instrument level is a precision instrument used to check and establish grades and elevations, and to set up level points over long distances. The following provides a review of the principal parts and setup of an instrument level:

- *Telescope* – The telescope on a typical instrument level sits on top of a bracelike device called the level bar. The level bar is essentially the body of the instrument level. The telescope itself consists of an eyepiece, located at the rear of the instrument, and an objective lens, located at the front of the instrument. An image-focusing screw knob near the center of the telescope is used to adjust the focus of the telescope to suit the user. Some telescopes are also equipped with adjustment screws that allow the eyepiece guide and crosshair ring to be adjusted. When the level is properly aligned onto a target, the image of the target should be aligned to both the vertical and horizontal cross hairs of the telescope.

  A sunshade is added to the front end of some telescopes to help reduce glare and to protect the objective lens. Dust caps are usually supplied with the telescopes to protect their objective lenses and, in some cases, a cover is also supplied to cover the eyepiece.

- *Spirit level* – The spirit level of an instrument level is mounted on top of the level bar that supports the telescope. The level tube is positioned horizontally parallel to the telescope. This spirit level is a very sensitive leveling device. It reflects (shows) the angular position of the level bar and telescope. If the front end of the telescope is lower than the rear, the bubble of the spirit level will be closer to the rear end of the level tube. The opposite is true if the telescope's front end is higher than the rear end. This spirit level is the level that must be watched when mounting the instrument level onto a tripod.

- *Leveling screws* – At the base of the instrument level is the foot plate, which is the part of the instrument that attaches to the mounting plate of the tripod. Just above the foot plate are three or four leveling screws that support the leveling head. The leveling head is actually the bottom end of the level bar. It may or may not have a horizontal graduated circle on it to indicate the horizontal angles at which the level is pointing. The leveling screws are used a set at a time to level the instrument before it is used.

Because the leveling screws used with an instrument level have limitations on how far they can be adjusted, it becomes even more critical that the tripod be set up as level as possible before the instrument is installed on top of the tripod's mounting plate. If the tripod is set up level and the leveling screws are adjusted to midrange, the instrument level should be easy to adjust so that it is level. When an instrument level is being set up and checked out before being used, the instrument's operator must make sure that the level reads level in all four directions. *Figure 24* shows a close-up view of the leveling screws used with an instrument level.

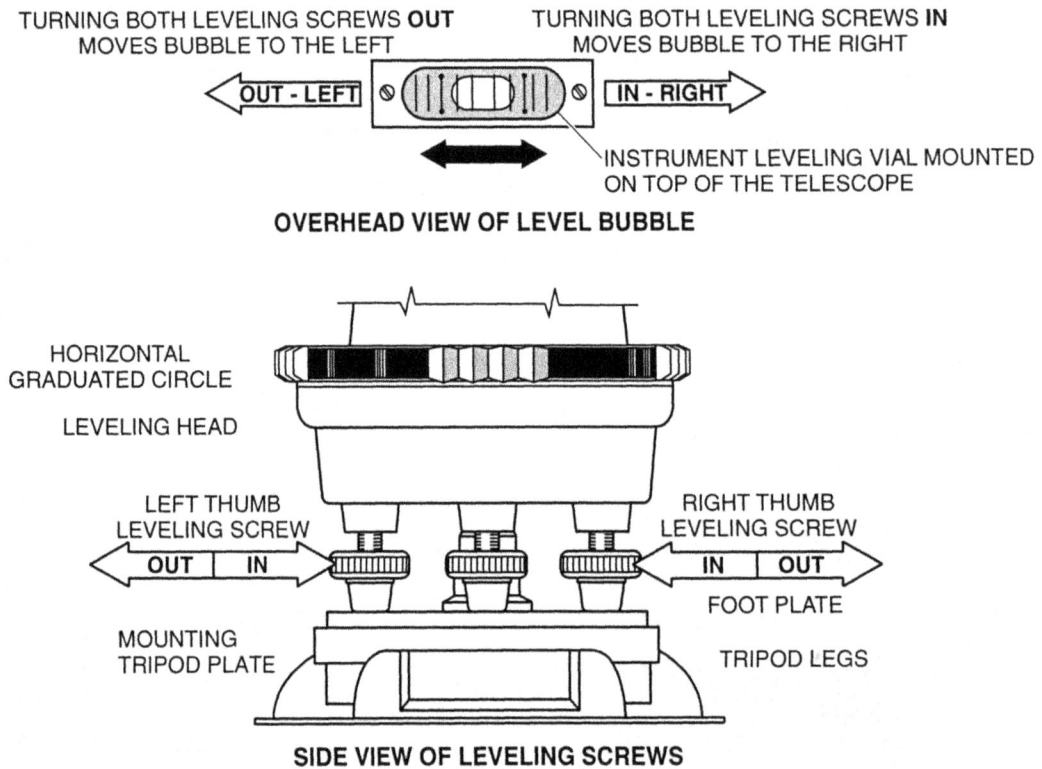

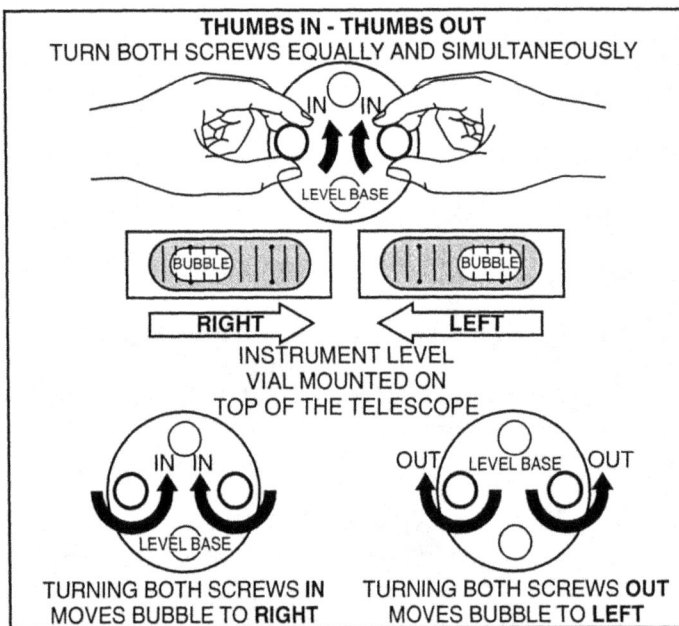

TURNING BOTH LEVELING SCREWS **OUT** MOVES BUBBLE TO THE LEFT

TURNING BOTH LEVELING SCREWS **IN** MOVES BUBBLE TO THE RIGHT

OUT - LEFT

IN - RIGHT

INSTRUMENT LEVELING VIAL MOUNTED ON TOP OF THE TELESCOPE

OVERHEAD VIEW OF LEVEL BUBBLE

HORIZONTAL GRADUATED CIRCLE

LEVELING HEAD

LEFT THUMB LEVELING SCREW

OUT   IN

RIGHT THUMB LEVELING SCREW

IN   OUT

FOOT PLATE

MOUNTING TRIPOD PLATE

TRIPOD LEGS

SIDE VIEW OF LEVELING SCREWS

THUMBS IN - THUMBS OUT
TURN BOTH SCREWS EQUALLY AND SIMULTANEOUSLY

IN   IN

LEVEL BASE

BUBBLE

BUBBLE

RIGHT

LEFT

INSTRUMENT LEVEL VIAL MOUNTED ON TOP OF THE TELESCOPE

IN   IN

LEVEL BASE

OUT   LEVEL BASE   OUT

TURNING BOTH SCREWS **IN** MOVES BUBBLE TO **RIGHT**

TURNING BOTH SCREWS **OUT** MOVES BUBBLE TO **LEFT**

103F24.EPS

Figure 24 ◆ Leveling screws used with an instrument level.

- *Horizontal clamps* – When the instrument level is positioned on top of the tripod's mounting plate, it is held in place by one or more horizontal clamp screws. When the clamp screws are tightened, they prevent the instrument from moving left to right on top of the tripod's mounting plate.
- *Horizontal tangent screw* – After an instrument level has been clamped to the tripod mounting plate, the only way to move the telescope left or right is with a horizontal tangent screw. On those instruments equipped with a horizontal tangent screw, located just below the level bar, the screw allows for very fine left and right adjustments to the telescope's horizontal position.
- *Chain hook* – Underneath the center of the instrument level's foot plate is a small chain hook that allows a plumb bob to be attached to the instrument. The plumb bob is suspended down through the hole in the center of the tripod's mounting plate.

The accuracy of an instrument level depends on its being properly set and leveled. Because levels and tripods differ in the way they are joined, this text will not address the specifics of mounting an instrument level onto a tripod. The following steps are performed on the assumption that the level is already on the tripod. Perform the following steps to set up and level an instrument level:

Step 1 Rotate the telescope horizontally so that it is directly over a pair of leveling screws (front and rear or left and right).

Step 2 Grasp the left and right leveling screws of one set of leveling screws between your thumb and forefinger.

> **NOTE**
> Leveling screws are turned in opposite directions, and they must be turned equally at the same time.

Step 3 Observe the bubble of the spirit level on the instrument, and then turn both screws in opposite directions (in or out) at the same time until the bubble is centered.

> **NOTE**
> On most instruments, the bubble moves in the same direction as your left thumb. Turning both screws in moves the bubble to the right. Turning both screws out moves the bubble to the left.

Step 4 Verify that the bubble has settled at the center of the spirit level before continuing.

Step 5 Rotate the telescope horizontally 90 degrees until it is aligned exactly over the remaining pair of leveling screws.

Step 6 Repeat Steps 1 through 4 until the bubble of the spirit level is centered.

Step 7 Rotate the telescope back to the first set of leveling screws, and make any minor leveling adjustments.

Step 8 Start at the 0° point on the horizontal graduated circle, and then rotate the telescope through each of the four leveling positions (quadrants) as a final check to make sure that the bubble stays centered:
- First quadrant is 0° to 90°.
- Second quadrant is 90° to 180°.
- Third quadrant is 180° to 270°.
- Fourth quadrant is 270° to 360° or 0°.

Step 9 Look across the top of the telescope, and aim the telescope at an object located away from the level.

Step 10 Look through the telescope at the targeted object, and adjust the focus knob on the telescope until the object is as clear as possible.

Step 11 Tighten the horizontal clamp to secure the level on the target, and then use the horizontal tangent screw to make any fine adjustments.

The process of differential leveling is based on the measurement of vertical distances from a level line. Elevations are transferred from one point to another by using a leveling instrument to first read a rod held vertically on a point of known elevation, then to read a rod held on a point of unknown elevation (*Figure 25*). Following this, the unknown elevation is calculated by adding or subtracting the readings. To determine elevations between two or more widely separated points or points on a sloping terrain, several repetitions of the same basic differential leveling process are performed.

7.1.0 Terminology Used in the Differential Leveling Process

Before describing the differential leveling process, it is important to first review and/or introduce some related terms:

- *Elevation* – The vertical distance above a datum point. For leveling purposes, a datum is normally based on the ocean's mean sea level (MSL). At numerous locations throughout the United States, the government has installed

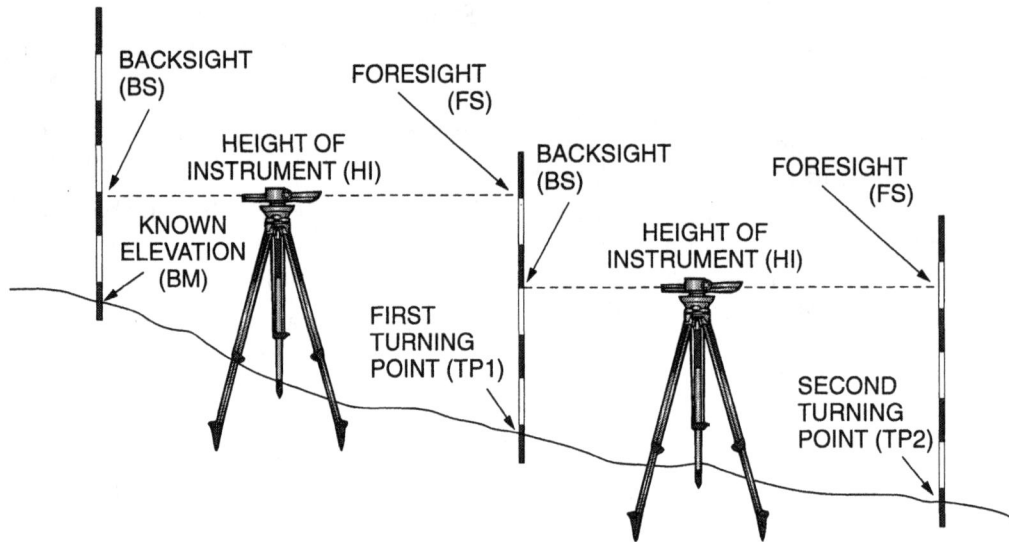

KNOWN ELEVATION (BM) + BACKSIGHT (BS) = HEIGHT OF INSTRUMENT (HI)
HEIGHT OF INSTRUMENT (HI) – FORESIGHT (FS) = TURNING POINT (TP) ELEVATION

103F25.EPS

Figure 25 ◆ Differential leveling relationships.

monuments marked with known elevations referenced to MSL. When readily available, such a monument would be used as the elevation reference. When no monument is readily available, such as at most construction sites, a point is established and an elevation arbitrarily assigned. Typically, an elevation of 100.00', 500.00', or 1,000.00' is used.

- *Bench mark (BM)* – A relatively permanent object with a known elevation located near or on a site. It can be iron stakes driven into the ground, a concrete monument with a brass disk in the middle, a chiseled mark at the top of a concrete curb, etc.

- *Backsight (BS)* – A reading taken on a leveling rod held on a point of known elevation to determine the height of the leveling instrument.

- *Foresight (FS)* – A reading taken on a leveling rod held on a point in order to determine the elevation.

- *Height of instrument (HI)* – The elevation of the line of sight of the telescope above the datum plane. It is determined by adding the backsight elevation to the known elevation.

- *Turning point (TP)* – A temporary point whose elevation is determined by differential leveling. The turning point elevation is determined by subtracting the foresight elevation from the height of the instrument elevation.

- *Closed loop* – A traverse consisting of a series of differential measurements that return to the point from which they began.

7.2.0 Differential Leveling Procedure

Before beginning the leveling process, you should select the bench mark that is closest to your work. If you do not know the exact location of the closest bench mark to your work, refer to the site plot plan. You should also determine the longest reasonable distance between your measurement points in order to shorten the amount of work you must do by minimizing instrument setups. Note that some sites are relatively flat, while others have steep slopes. Regardless of the slope involved, the procedure for leveling is the same. The difference is that when leveling at a site with a very steep slope, the procedure becomes more time consuming. This is because the line of sight of the instrument relative to intercepting a leveling rod is shorter, requiring that more setups be used to cover the distance involved.

The differential leveling procedure generally involves two people working together and communicating with each other. One person is designated as the rod person and the other the instrument person.

Depending on the complexity of the task, recording of the collected measurement data in the field notes may be done by either person, both, or sometimes by a third person. An example of a typical differential elevation procedure is described next, and its path (traverse) is shown in *Figure 26*.

Step 1 The procedure begins by recording the starting point (BM) and its known elevation in the **station** and elevation columns of the field notes. For the example shown, the entries are BM (station) and 1,000.00' (elevation).

Step 2   The instrument person sets up the leveling instrument at Station 1 (STA 1) in preparation for the first measurement. It should be located so that a level rod placed on the BM is in the line of sight of the level and the rod can be clearly read. Note that this same point should also allow the line of sight of the level to intercept a level rod held on the proposed location of the first turning point (TP1). Set this point equally distant between the two points and no farther away than 150' to 200' from either point of measurement.

Step 3   While the rod person holds the level rod plumb on the BM, a backsight rod reading is taken, then recorded in the field notes. For our example, the BS reading of 7.77' is recorded in the BS (+) column of the notes. Following this, the height of the instrument (HI) is calculated and recorded in the HI column of the field notes. For our example, the HI is recorded as 1,007.77' (HI = BM + BS = 1,000.00 + 7.77').

Step 4   The rod person paces or otherwise measures the approximate distance between the BM and the leveling instrument and then advances an equal distance beyond the level in the desired direction of the first turning point (TP1). This point must be located such that when the level rod is placed on it, the line of sight of the leveling instrument will intercept the rod. The rod person selects an appropriate solid surface, such as a sidewalk or large rock, for the turning point. Note that an unmarked point on grass or soil should never be used as a turning point. If no natural solid object is available, a metal turning pin, railroad spike, or wooden stake driven in the ground can serve as a turning point. When a turning point on a solid surface such as a sidewalk or pavement is used, the point should be marked and identified by the turning point number.

Step 5   While the rod person holds the level rod plumb on TP1, a foresight rod reading is taken, then recorded in the field notes. For our example, the FS reading of 5.23' is recorded in the FS (−) column. Following this, the elevation of TP1 is calculated and recorded in the elevation column of the field notes. For our example, the elevation is recorded as 1,002.54' (turning point elevation = HI − FS = 1,007.77' − 5.23').

Step 6   In preparation for the next set of backsight and foresight readings, the instrument person moves the leveling instrument to a point beyond TP1 and sets up the instrument at Station 2, which is approximately midway between TP1 and TP2.

Step 7   Once the leveling instrument is set up, backsight and foresight readings are taken between the points TP1 and TP2 in the same way as previously described in Steps 3 through 5, with the following exceptions. The known elevation of TP1 is used instead of the BM to calculate the instrument height (HI) at Station 2. Then, the new HI and the foresight reading on TP2 are used to calculate the elevation of TP2.

Step 8   Steps 3 through 7 are repeated as necessary to complete the differential measurement loop from the TP2 to the **temporary bench mark** TBM1, then back via TP3 to the starting point at BM.

In the example shown, the leveling traverse is run back to the starting point at BM. This is called closing the loop or a closed level loop. Any leveling survey should close back either on the starting bench mark or on some other point of known elevation in order to provide a check of the measurements taken.

Leveling notes should always be checked for arithmetic or calculator input errors. This is done by simply summing the backsight (BS) and foresight (FS) columns and comparing the difference between them with the starting and ending elevations. As shown in *Figure 26*, the difference between the BS sum and the FS sum is 0.00'. Also, the difference between the starting elevation of 1,000.00' and the ending elevation of 1,000.00' is 0.00'. Since the differences are equal, the arithmetic checks and the loop is properly closed. An error would exist if the differences were not equal or were not within the established accuracy standard or tolerances specified for the project. Using the same example, the calculations for the traverse between the BM and TBM1 can be checked in the same manner. This is shown in *Figure 26*.

When performing differential leveling, it is easy to make mistakes. However, mistakes can be eliminated by constantly checking and rechecking your work. The following are some common mistakes to avoid when performing differential leveling:

- Using unequal backsight and foresight distances
- Using an instrument that is not level
- Not ensuring that the rod is plumb (if not using a level, the rod should be rocked forward and backward, then the smallest reading recorded)

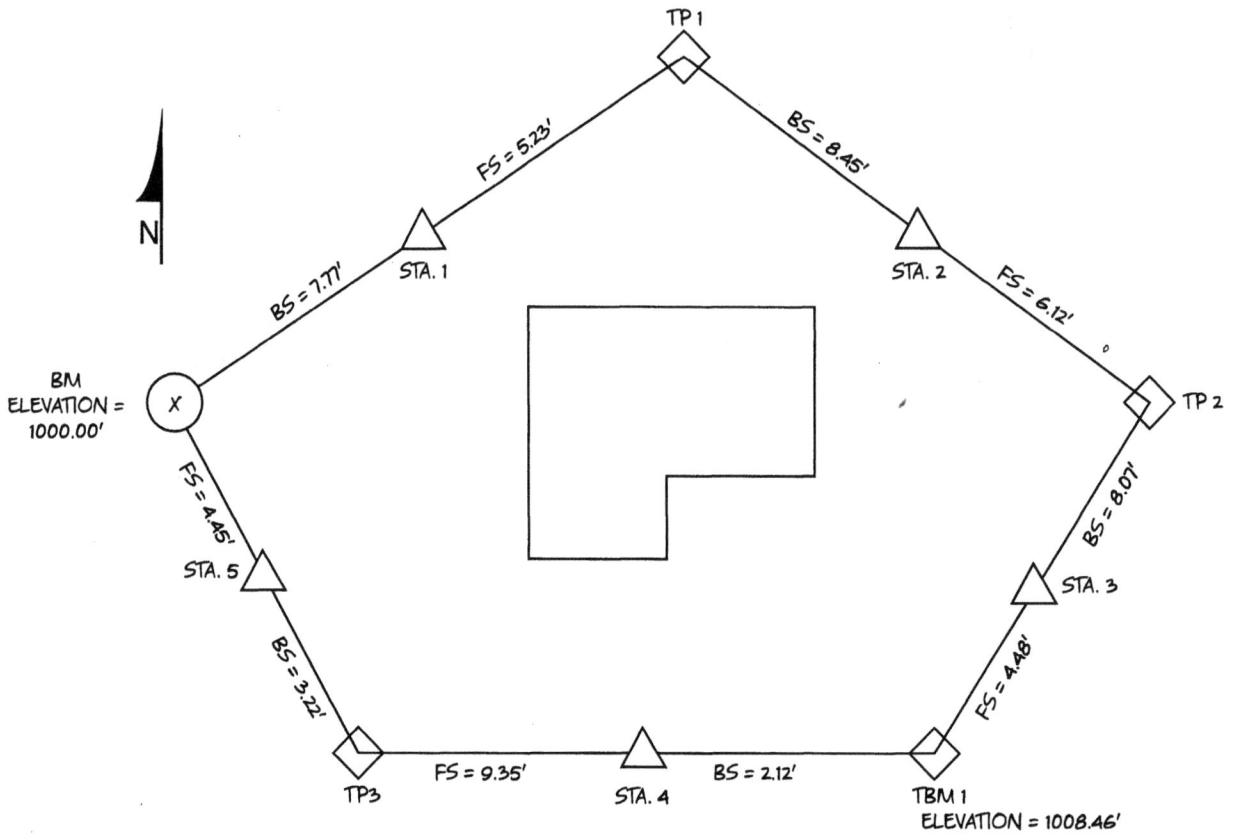

STATION (STA)	BS (+)	HI	FS (−)	ELEVATION
BM				1000.00'
STA. 1	7.77'	1007.77'		
TP 1			5.23'	1002.54'
STA. 2	8.45'	1010.99'		
TP 2			6.12'	1004.87'
STA. 3	8.07'	1012.94'		
TBM 1			4.48'	1008.46'
STA. 4	2.12'	1010.58'		
TP 3			9.35'	1001.23'
STA. 5	3.22'	1004.45'		
BM			4.45'	1000.00'
Σ CHECK	29.63'		29.63'	

DIFFERENCE = 0.00'

DIFFERENCE = 0.00'

MATH CHECK:

```
  1000.00
+   24.29
  1024.29
-   15.83
  1008.46
```

103F26.EPS

Figure 26 ◆ Differential leveling traverse and related field notes data.

- Measuring when sections of an extended leveling rod are not adjusted properly
- Misreading the rod
- Letting dirt, ice, or other debris accumulate on the base of the rod
- Recording incorrect values in the field notes
- Moving the position of a turning point between backsight and foresight readings

8.0.0 ◆ FIELD NOTES

In the taping and differential leveling procedures described in the module *Introduction to Site Layout*, constant reference was made to recording measurement information in field notes. Writing a legible and accurate set of notes in a field book (*Figure 27*) is just as important as doing the leveling or layout work itself because field notes provide a historical record of the work performed. They serve as a reference if there is a question about the correctness or integrity of your work, especially in a court of law. Field notes should leave no room for misinterpretation. Your notes should be written so that others can understand your work. General guidelines for writing and keeping field notes follow:

- All field notebooks should contain the name, address, and phone number of the owner.
- All pages should be numbered, and there should be a table of contents page.

- Make neatly printed entries in the book using a suitable sharp pencil with hard lead (3H or 4H). Never use cursive script in a field book.
- Begin each new task on a new page. The left-hand pages are generally used for entering numerical data, and the right-hand pages are for making sketches and notes.
- Always record the date, time, weather conditions, names of crew members and their assignments, and a list of the equipment used.
- Record each measurement in the field book immediately after it is taken. Do not trust it to memory.
- Record data exactly. Ideally, the data should be checked by two crew members at the time it is recorded.
- Make liberal use of sketches if needed for clarity. They should be neat and clearly labeled, including the approximate north direction. Do not crowd the sketches.
- Never erase. If a mistake is made, draw a single line through the incorrect entry and write the correct data above it.
- Draw a diagonal across the page and mark the word VOID on the tops of pages that, for one reason or another, are invalid. When marking the page, be careful not to make the voided information unreadable. The date and name of the person voiding the page should also be recorded.

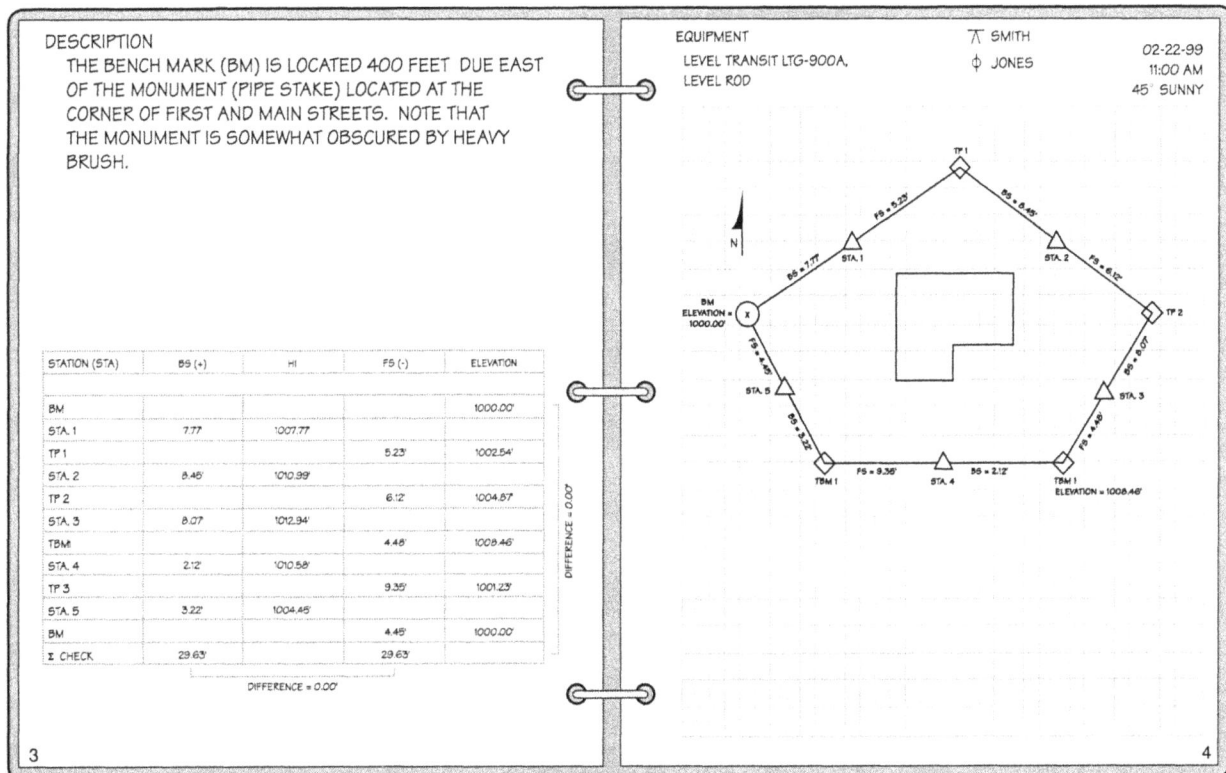

DESCRIPTION
THE BENCH MARK (BM) IS LOCATED 400 FEET DUE EAST OF THE MONUMENT (PIPE STAKE) LOCATED AT THE CORNER OF FIRST AND MAIN STREETS. NOTE THAT THE MONUMENT IS SOMEWHAT OBSCURED BY HEAVY BRUSH.

STATION (STA)	BS (+)	HI	FS (-)	ELEVATION
BM				1000.00'
STA. 1	7.77'	1007.77'		
TP 1			5.23'	1002.54'
STA. 2	8.45'	1010.99'		
TP 2			6.12'	1004.87'
STA. 3	8.07'	1012.94'		
TBM			4.48'	1008.46'
STA. 4	2.12'	1010.58'		
TP 3			9.35'	1001.23'
STA. 5	3.22'	1004.45'		
BM			4.45'	1000.00'
Σ CHECK	29.63'		29.63'	

DIFFERENCE = 0.00'

EQUIPMENT
LEVEL TRANSIT LTG-900A,
LEVEL ROD

人 SMITH
φ JONES

02-22-99
11:00 AM
45° SUNNY

3

4

103F27.EPS

Figure 27 ◆ Notes in a field book.

- Mark the word COPY on the top of copied pages. Refer to the name and page number of the original document.
- Always keep the field book in a safe place on the job site. At night, lock it up in a fireproof safe. Original field books should never be destroyed, even if copied for one reason or another.

9.0.0 ◆ BATTER BOARDS

On some construction jobs, wooden frameworks called batter boards (*Figure 28*) are used to establish building and other construction layout lines. Used in pairs and with a string or wire attached and stretched between them, batter boards are used to create lines that mark the boundaries of a building, the center of column footings, etc. They can also be used to set reference elevations, such as elevations to the top of a footing or to the finish floor level of a building.

A typical batter board consists of a 2 × 4 or 2 × 6 horizontal board, called a ledger board, that is nailed

or otherwise attached to stakes driven into the ground. Typically, the stakes are made from 2 × 4s.

> **NOTE**
>
> In locations where the ground is very hard, the stakes used to support ledger boards are often made from rebar. This is because rebar stakes are much easier to drive into the ground than wooden stakes.

The placement of batter boards is normally done after the exact locations of the building corners have been established. Placement involves driving the ledger board support stakes firmly into the ground behind each building corner at a distance that allows enough working room between the batter boards and the immediate construction area. Depending on the type of job and the excavation equipment used, this distance could be anywhere between 4' and 20'. If it is necessary to drive

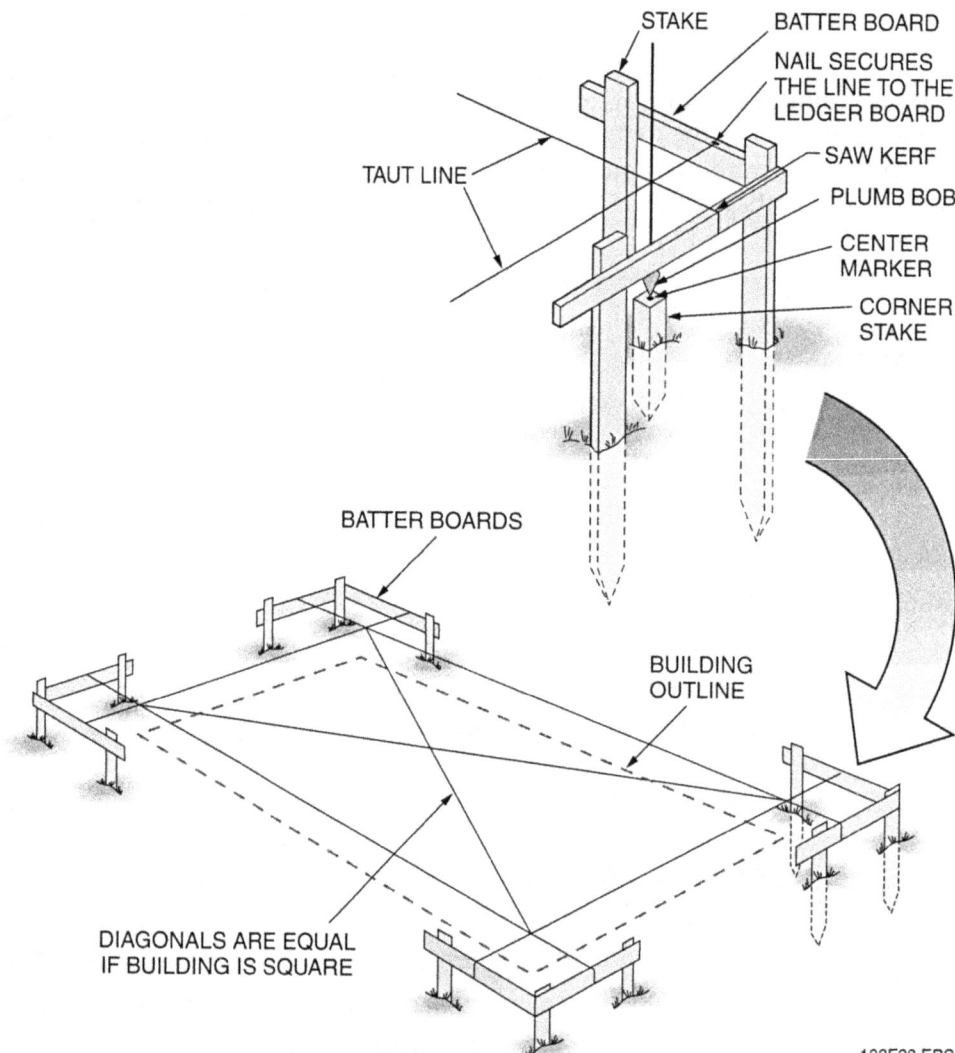

Figure 28 ◆ Typical use of batter boards.

103F28.EPS

SITE LAYOUT LEVEL ONE — TRAINEE MODULE 78103-04

the stakes in soft soil, or if the stakes extend 3' or more out of the ground, they should be braced to prevent any movement. Following this, a leveling instrument is used to sight and mark the stakes at the required elevation. Then, each ledger board is fastened to the outside of its support stakes so that its top edge is on the elevation mark. It is important to make sure that when all the related batter boards have been installed, the tops of the ledger boards are level with one another.

Once the batter boards are installed, the building corners can be transferred to the batter boards. One method for doing this involves the use of a plumb bob and nylon string. This is done by stretching a nylon string (line) between two opposite batter boards and directly over the building corner stakes. The plumb bob is used to locate the exact position of the line by suspending it directly over the center marker on each corner stake. When the line is accurately located on the two batter boards, a shallow saw cut (kerf) is made at this point on the outside top edge of each ledger board. This prevents the line from moving when stretched and secured. The taut line is placed in the kerf and secured with a nail driven on the back of each ledger board. The procedure is repeated until all the building lines are in place.

After all the lines are installed between the batter boards, measurements should be made between the lines to make sure that they are accurate. Also, the diagonals across the lines should be measured to make sure that they are equal. Equal-length diagonals indicate that the lines are square.

10.0.0 ◆ LAYING OUT AND CHECKING 90° ANGLES USING THE 3-4-5 RULE

The 3-4-5 rule has been used in construction for centuries. It is a simple method that can be used for laying out or checking a 90° angle that does not require the use of a builder's level or transit. The numbers 3-4-5 represent dimensions in feet that describe the sides of a right triangle. The 3-4-5 rule is based on the Pythagorean theorem, which was discussed in detail in the module *Surveying Math*. It states that in any right triangle, the square of the longest side, called the hypotenuse (C), is equal to the sum of the squares of the two shorter sides (A and B). Stated mathematically:

$$C^2 = A^2 + B^2$$

Accordingly, for the 3-4-5 right triangle:

$$5^2 = 3^2 + 4^2$$

$$25 = 9 + 16$$

$$25 = 25$$

This theorem also applies if you multiply each number (3, 4, and 5) by the same number. For example, if multiplied by the constant 3, it becomes a 9-12-15 triangle.

For most construction layout and checking, right triangles that are multiples of the 3-4-5 triangle are used, such as 9-12-15, 12-16-20, 15-20-25, and 30-40-50. The specific multiple used is determined mainly by the relative distances involved in the job being laid out or checked. It is best to use the highest multiple that is practical. When smaller multiples are used, any error made in measurement will result in a much greater angular error.

Figure 29 shows an example of the 3-4-5 rule involving the multiple 48-64-80. In order to square or check a corner as shown in the example, first measure 48'-0" down the line in one direction, then 64'-0" down the line in the other direction. The distance measured between the 48'-0" and 64'-0" points must be exactly 80'-0" if the angle is to be a perfect right angle. If the measurement is not exactly 80', the angle is not 90°. This means that the direction of one of the lines or the corner point must be adjusted until a right angle exists.

It cannot be emphasized enough that exact measurements are necessary to get the desired results when using the 3-4-5 method of laying out or checking a 90° angle. Any error in the measurements of the distances will result in not establishing a right angle. If an existing 90° angle is being checked, inaccurate measurements may cause you to make an unnecessary adjustment.

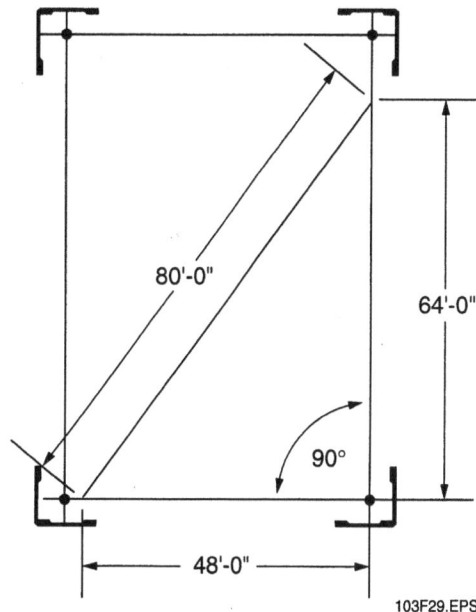

103F29.EPS

Figure 29 ◆ Checking lines for square using the 48-64-80 multiple of a 3-4-5 triangle.

11.0.0 ◆ COMMUNICATING WITH HAND SIGNALS

Two-way radios are commonly used to maintain voice communication between members of a site layout crew. However, because of equipment noises, blasting, non-availability of radios, or other reasons, the use of hand signals may be required. *Figure 30* shows some common hand signals for use when performing distance measurement and differential leveling tasks.

MOVE IN THIS DIRECTION

MOVE UP

MOVE DOWN

TURNING POINT

WAVE ROD FROM SIDE TO SIDE; ROCK BACK AND FORTH

OBSERVATION COMPLETED; MOVE ON; UNDERSTAND

WRONG FACE; CHECK CLAMP; ROD UPSIDE DOWN

USE LONG ROD

COME IN

103F30.EPS

Figure 30 ◆ Common hand signals.

Summary

Site preparation and layout tasks can be performed by different individuals, including surveyors, field engineers, and trainees under the supervision of surveyors and field engineers. However, because of the tremendous liability involved, only professional load surveyors should make any measurements that relate to property lines and boundaries.

It is important to eliminate mistakes and reduce the size of errors in measurement. This is achieved by having a good understanding of measurement principles and by rechecking your work several times. It is also achieved by making sure to always use calibrated measuring instruments. Do not use instruments that are damaged or otherwise incapable of measuring within tolerances. Fundamental rules for handling all instruments and tools used for site layout include the following:

- Only use equipment that you know how to operate.
- Do not force a part if it is difficult to move.
- Keep equipment clean.
- Use protective cases when provided.
- When an instrument is wet, let it dry naturally, not closed up in its case.
- Do not leave equipment unattended.

A variety of instruments, including the transit, theodolite, and total station, can be used to establish and measure horizontal and vertical angles. Some common mistakes made when establishing or measuring angles include the following:

- Poor setup and leveling of the instrument
- Misreading the instrument scale indications
- Transposing and/or recording the wrong angle values in field notes
- Sighting on the wrong targets, marks, or lines when measuring horizontal or vertical angles
- Using the wrong instrument tangent screw
- Failure to center the telescope bubble before measuring vertical angles
- Failure to take into account the algebraic sign for the values of vertical angles measured with a transit

In differential leveling, elevations are transferred from one point to another by using a leveling instrument to first read a rod held vertically on a point of known elevation (backsight) in order to determine the height of the instrument (HI), then a rod held on a point of unknown elevation is read (foresight). Following this, the unknown elevation is calculated by adding or subtracting the foresight reading from the HI.

Some common mistakes made when leveling include the following:

- Not setting up the instrument on a stable surface
- Not making sure that the instrument is level
- Not holding the leveling rod in a plumb position
- Not fully extending the rod
- Reading the rod incorrectly
- Setting the wrong datum elevation
- Not closing the loop

Writing a legible and accurate set of notes in a field book is just as important as doing the leveling or layout work itself. This is because field notes provide a historical record of the work performed. They serve as a reference should there ever be a question about the correctness or integrity of your work, especially in a court of law. Some common mistakes made in field notes include the following:

- Mathematical errors
- Incorrect data
- Illegible data
- Erasures

Review Questions

1. An automatic level is designed to perform the same functions as a transit.
   a. True
   b. False

Refer to *Figure 1* when answering Questions 2 and 3.

2. The instrument labeled B in *Figure 1* is a(n) _____.
   a. theodolite
   b. transit
   c. total station
   d. electronic distance measurement instrument (EDMI)

3. The instrument labeled D in *Figure 1* is a(n) _____.
   a. total station
   b. transit
   c. electronic theodolite
   d. rotating laser

4. Of the transit and the theodolite, the _____ generally provides more precise readings.
   a. transit
   b. theodolite

5. All laser instruments transmit a(n) _____ beam.
   a. infrared
   b. visible light
   c. invisible light
   d. coherent light

(A)

(B)

(C)

(D)

(E)

103RQ01.EPS

6. All of the following are rules for using a laser instrument *except* _____.
   a. wear approved safety goggles when the laser has a power output of 5 milliwatts or greater
   b. avoid going into areas marked off for laser operation unless it is necessary because of the job
   c. only employees who have carefully read the manufacturer's operating instructions for the instrument are permitted to set up and operate a laser
   d. never point a laser beam at anyone

Refer to *Figure 2* when answering Question 7.

7. What is the clockwise angular reading indicated by the vernier shown in *Figure 2*?
   a. 15°25'00"
   b. 17°25'00"
   c. 340°05'00"
   d. 342°35'00"

8. An optical plummet is used to _____.
   a. set an instrument up over a reference point
   b. align an instrument at 0° in the horizontal plane
   c. measure the height of an instrument
   d. reflect a laser beam back to its source

9. The purpose of a horizontal crosshair test is to verify that the instrument's horizontal crosshair is in a plane that is _____ to the vertical axis of the instrument.
   a. perpendicular
   b. parallel

10. A telescope bubble test is performed to verify that the _____.
    a. instrument's line of sight is parallel to the horizontal axis
    b. instrument's horizontal axis is perpendicular to the vertical axis
    c. axis of the telescope's level is parallel to its line of sight
    d. vertical circle reads zero when the instrument is properly leveled and the telescope bubble is centered

11. When using a standard transit to turn a foresight reading for a horizontal angular measurement, the upper motion clamp should be _____ and the lower motion clamp should be _____.
    a. loosened; loosened
    b. loosened; tightened
    c. tightened; tightened
    d. tightened; loosened

12. When measuring a 65°00'00" angle by repetition, what would you expect the instrument scale to read for the accumulated value of the angle after making eight repetitions?
    a. 30°00'00"
    b. 65°00'00"
    c. 130°00'00"
    d. 160°00'00"

13. Once the horizontal clamps on an instrument level are tightened, it is still possible to move the instrument left or right using the horizontal tangent screw.
    a. True
    b. False

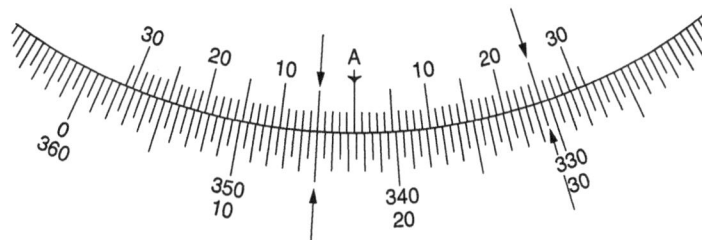

103RQ02.EPS

Figure 2

14. In the differential leveling procedure, the _____ is determined by _____ the backsight elevation _____ the known elevation.
    a. height of instrument (HI); adding; to
    b. turning point (TP) elevation; subtracting; from
    c. height of instrument (HI); subtracting; from
    d. turning point (TP) elevation; adding; to

Refer to *Figure 3* when answering Question 15.

15. What is the elevation of the anchor bolt (A-bolt) shown in *Figure 3*?
    a. 781.22'
    b. 784.99'
    c. 787.04'
    d. 792.86'

16. Each of the entries below is a recommended practice for making field notes *except* _____.
    a. sketches should be used freely to illustrate your work
    b. if it is necessary to correct any measurement data, neatly erase it and write in the correct data
    c. make references to any information that has been copied from another source
    d. write notes so that they leave no room for misinterpretation

17. When setting up batter boards, the ledger board is positioned so that _____.
    a. it is plumb
    b. the top edge of the board is level with the elevation mark
    c. it is within one foot of the building corner
    d. it is at least 3' above the ground

18. When checking the angle formed by the corner of a 15' wall and a 20' wall for square using the 3-4-5 rule, the best angular accuracy is obtained if the multiple of _____ is used.
    a. 9-12-15
    b. 12-16-20
    c. 15-20-25
    d. 30-40-50

19. If a foundation measures 30' wide and 40' long, the diagonal should be _____.
    a. 45'
    b. 50'
    c. 55'
    d. 60'

20. The hand signal in which the arm is held straight up with the palm of the hand facing out represents _____.
    a. stop; observation complete
    b. move up
    c. come in
    d. use long rod

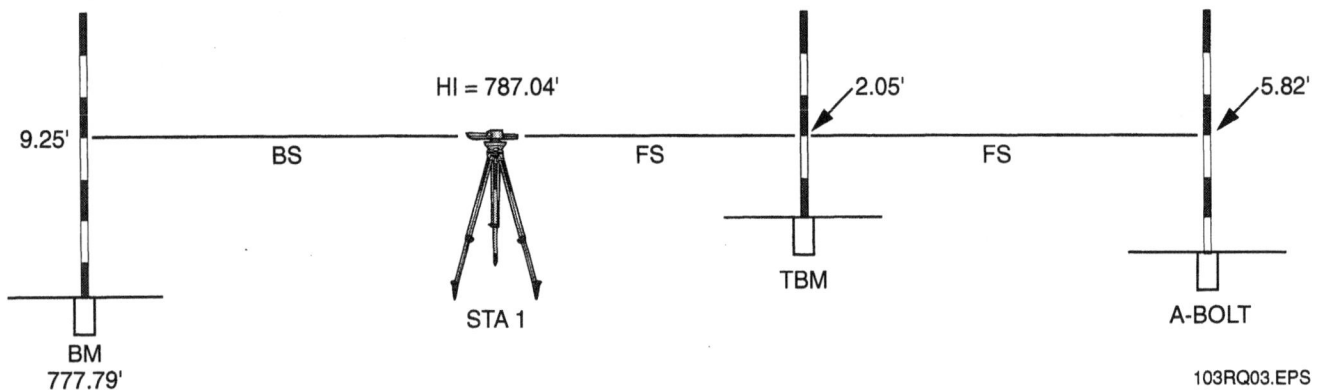

Figure 3

Trade Terms Introduced in This Module

Alidade: The upper part of a transit or theodolite that contains the telescope and related components.

Axis: A straight reference line, either real or imaginary, passing through an object, around which the object revolves or around which parts of the object are symmetrically arranged.

Differential leveling: A method of leveling used to determine the difference in elevation between two points.

Field notes: A permanent record of field measurement data and related information.

Height of instrument (HI): The elevation of the line of sight of the telescope relative to a known elevation. It is determined by adding the backsight elevation to the known elevation.

Laser instrument: A device that transmits a very narrow, intense beam of coherent light. *Laser* is an acronym for light amplification by stimulated emission of radiation.

Least count: The finest reading that can be made directly on a vernier of a transit or micrometer of a theodolite.

Optical plummet: A device incorporated into a transit, theodolite, or similar instrument that allows the operator to sight to a point below that is exactly plumb with the center of the instrument. This enables quick and accurate setup of the instrument over a point.

Parallax: The apparent movement of the crosshairs in a surveying instrument caused by movement of the eyes.

Plunge the telescope: To reverse the direction of an instrument telescope around its horizontal axis.

Sexagesimal system: A system used in surveying in which the circumference of a circle is divided into 360 parts or degrees. Each degree is further divided into minutes and seconds, where one degree is equal to 60 minutes and one minute is equal to 60 seconds.

Slope: A measurement of how much the ground varies from horizontal.

Station: Instrument setting location in differential leveling.

Temporary bench mark: A point of known (reference) elevation determined from bench marks through leveling. The temporary bench mark must be permanent enough to last for the duration of a project.

Turning point (TP): A temporary point within an open or closed differential leveling circuit whose elevation is determined by differential leveling. It is normally the leveling rod location. Its elevation is determined by subtracting the foresight elevation from the height of the instrument elevation.

Vernier: A short auxiliary scale set parallel to a primary scale.

Vertical angle: An angle in the vertical plane measured up (+) or down (–) from horizontal.

Additional Resources

This module is intended to be a thorough resource for task training. The following reference works are suggested for further study. These are optional materials for continued education rather than for task training.

Construction Surveying and Layout: A Step-by-Step Engineering Methods Manual, 1995. Wesley G. Crawford. West Lafayette, IN: Creative Construction Publishing.

Principles and Practices of Commercial Construction, 2001. Cameron K. Andres and Ronald C. Smith. Upper Saddle River, NJ: Prentice Hall.

Surveying, 1999. Jack McCormack. New York, NY: John Wiley & Sons.

Surveying Practice, 1998. Jerry A. Nathanson, et al. New York, NY: McGraw-Hill, Inc.

Surveying Principles and Applications, 2000. Barry F. Kavanagh and Glen Bird. Upper Saddle River, NJ: Prentice Hall.

Surveying With Construction Applications, 1997. Barry F. Kavanagh. Upper Saddle River, NJ: Prentice Hall.

Figure Credits

John Hoerlein	103F01, 103F04, 103F05, 103RQ01(C), 103RQ01(D), 103RQ01(E)
Sokkia Corporation	103F02, 103F07 103RQ01B
Pacific Laser Systems	103F08
David White	103F09, 103F11, 103RQ01A

The NCCER makes every effort to keep these textbooks up-to-date and free of technical errors. We appreciate your help in this process. If you have an idea for improving this textbook, or if you find an error, a typographical mistake, or an inaccuracy in NCCER's Contren® textbooks, please write us, using this form or a photocopy. Be sure to include the exact module number, page number, a detailed description, and the correction, if applicable. Your input will be brought to the attention of the Technical Review Committee. Thank you for your assistance.

Instructors – If you found that additional materials were necessary in order to teach this module effectively, please let us know so that we may include them in the Equipment/Materials list in the Instructor's Guide.

Write: Product Development
National Center for Construction Education and Research
P.O. Box 141104, Gainesville, FL 32614-1104

Fax: 352-334-0932

E-mail: curriculum@nccer.org

Craft _____ Module Name _____

Copyright Date _____ Module Number _____ Page Number(s) _____

Description _____

_____

_____

_____

(Optional) Correction _____

_____

_____

(Optional) Your Name and Address _____

_____

_____

Blueprint Reading
for Surveyors

COURSE MAP

This course map shows all of the modules in the first level of the Site Layout curriculum. The suggested training order begins at the bottom and proceeds up. Skill levels increase as you advance on the course map. The local Training Program Sponsor may adjust the training order.

SITE LAYOUT LEVEL ONE

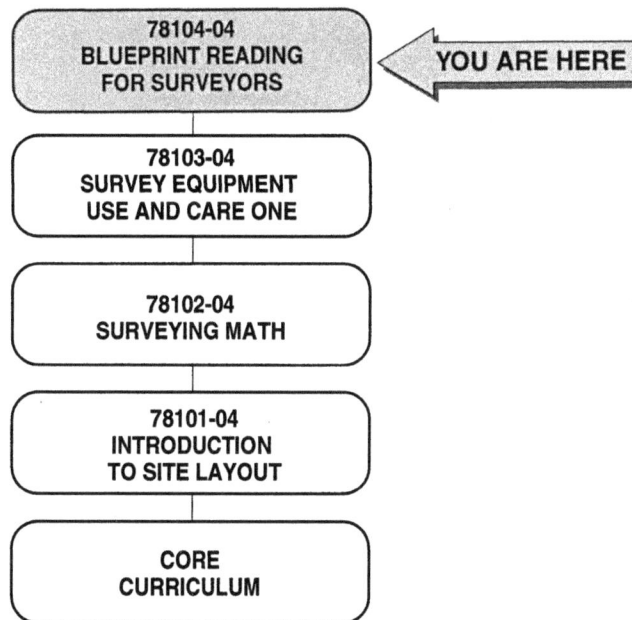

```
┌─────────────────────────┐
│ 78104-04                │        ◄── YOU ARE HERE
│ BLUEPRINT READING       │
│ FOR SURVEYORS           │
└─────────────────────────┘
┌─────────────────────────┐
│ 78103-04                │
│ SURVEY EQUIPMENT        │
│ USE AND CARE ONE        │
└─────────────────────────┘
┌─────────────────────────┐
│ 78102-04                │
│ SURVEYING MATH          │
└─────────────────────────┘
┌─────────────────────────┐
│ 78101-04                │
│ INTRODUCTION            │
│ TO SITE LAYOUT          │
└─────────────────────────┘
┌─────────────────────────┐
│ CORE                    │
│ CURRICULUM              │
└─────────────────────────┘
```

104CMAP.EPS

Figures

Tables

Blueprint Reading for Surveyors

Objectives

When you have completed this module, you will be able to do the following:

1. Describe the types of drawings usually included in a set of plans and list the information found on each type.
2. Identify the different types of lines used on construction drawings.
3. Identify selected architectural symbols commonly used to represent materials on plans.
4. Identify selected electrical, mechanical, and plumbing symbols commonly used on plans.
5. Identify selected abbreviations commonly used on plans.
6. Read and interpret plans, elevations, schedules, sections, and details contained in basic construction drawings.
7. State the purpose of written specifications.
8. Identify and describe the parts of a specification.
9. Demonstrate or describe how to perform a quantity takeoff for materials.
10. Read and interpret orthographic projection and isometric drawings.
11. Perform a quantity takeoff to determine a concrete quantity.
12. Calculate rebar required for an identified segment of a drawing.

Prerequisites

Before you begin this module, it is recommended that you successfully complete the following: Core Curriculum; Site Layout Level One, Modules 78101-04 through 78103-04.

1.0.0 ◆ INTRODUCTION

This module reviews and builds on the construction drawing (blueprint) material introduced in the *Core Curriculum*. It introduces techniques for reading construction drawings and specifications. Construction drawings tell how to build a specific building or structure. A specification is a contractual document used along with the construction drawings. It contains detailed written instructions that supplement the set of drawings. It is extremely important that you interpret construction drawings and specifications correctly. Failure to do so may result in costly rework and unhappy customers. Depending on the severity of the mistake, it can also expose you and your employer to legal liability.

2.0.0 ◆ DRAWING SET

The set of detailed drawings or plans drawn to scale by an architect and/or engineer shows all of the information and dimensions necessary to build or remodel a structure. Copies of the architect's original drawings are made for contractors and others to use. These copies are called blueprints. The term blueprint is derived from a method of reproduction that was used in the past. A true blueprint shows the details of the structure as white lines on a blue background. Today, most drawing reproduction methods produce a black, blue, or brownish-colored line on a white background. These copies or prints are typically made using a diazo or similar copying machine. However, the term blueprint is still widely used when referring to any copies made from original drawings.

Construction drawings or architect's plans consist of several different kinds of drawings assembled into a set (*Figure 1*). In order for complete information about a structure to be conveyed to the reader, various drawings in the set illustrate the structure using a variety of different views.

A set of drawings also includes sheets that contain relevant written information, such as notes and equipment or material schedules. Notes often contain important information. It is very important to read the notes.

The types of written information and views normally contained in a drawing set include the following:

- Title sheets, title blocks, and revision blocks
- Architectural drawings consisting of the following:
  - **Plan views**
  - Elevations
  - Sections
  - Details
  - Schedules
- Structural drawings
- Plumbing plans
- Mechanical plans
- Electrical plans

2.1.0 Title Sheets, Title Blocks, and Revision Blocks

A title sheet is normally placed at the beginning of a set of drawings or at the beginning of a major section of drawings. It provides an index to the other drawings; a list of abbreviations used on the drawings and their meanings; a list of symbols used on the drawings and their meanings; and various other project data, such as the project location, the size of the land parcel, and the building size. It is important that you use the title sheet(s) that come with the drawing set so that you understand the specific symbols and abbreviations used throughout the drawings. These symbols and abbreviations may vary from plan to plan.

A title block or box (*Figure 2*) is normally placed on each sheet in a set of drawings. It is usually located in the bottom right-hand corner of the sheet, but this location can vary.

The title block serves several purposes in terms of communicating information. Generally, it contains the name of the firm that prepared the drawings, the owner's name, and the address and name of the project. It also gives locator information, such as the title of the sheet, the drawing or sheet number, the date the sheet was prepared, the scale, and the initials or names of the people who prepared and checked the drawing.

A revision block is normally shown on each sheet in a set of drawings. Typically, it is located in the upper right-hand corner or bottom right-hand corner of the drawing, near or within the title block. It is used to record any changes (revisions) to the drawing. An entry in the revision block usually contains the revision number or letter, a brief description of the change, the date, and the initials of the person making the revision(s). When using drawings, it is essential to note the revision designation on each drawing and use only the latest issue; otherwise, costly mistakes will result. If in doubt about the revision status of a drawing, check with your supervisor to make sure that you are using the most recent version of the drawing. Also, check to see if a **request for information (RFI)** or sketches have been issued in the latest plan revision.

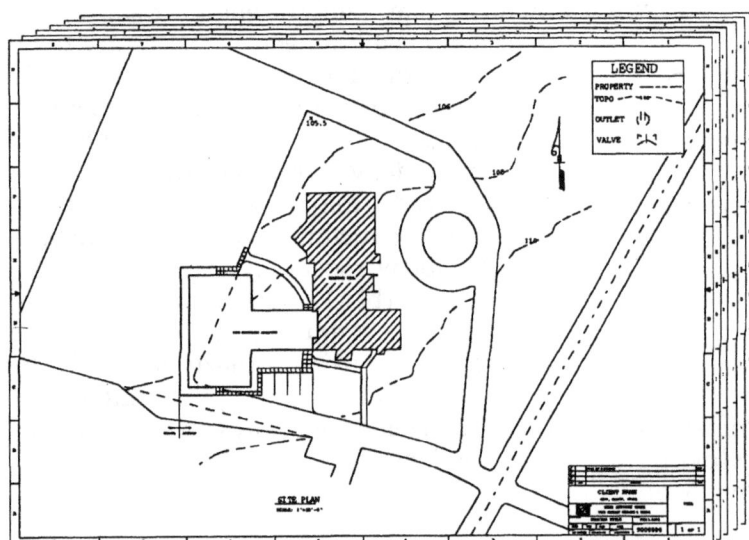

TITLE SHEET(S)
ARCHITECTURAL DRAWINGS
- SITE (PLOT) PLAN
- FOUNDATION PLAN
- FLOOR PLANS
- INTERIOR/EXTERIOR ELEVATIONS
- SECTIONS
- DETAILS
- SCHEDULES

STRUCTURAL DRAWINGS

PLUMBING PLANS

MECHANICAL PLANS

ELECTRICAL PLANS

104F01.TIF

Figure 1 ◆ Format of a working drawing set.

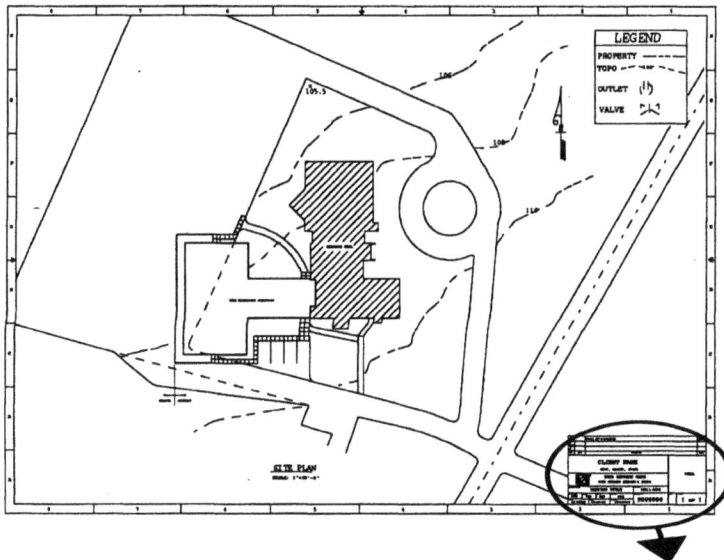

Figure 2 ◆ Title and revision blocks.

2.2.0 Plan View Drawings

Plan view drawings are drawings that show the structure looking down from above. The object is projected from a horizontal plane. Typically, plan view drawings are made to show the overall construction site (plot or site plan), the structure's foundation (foundation plan), and the structure's floor plans.

2.2.1 Site Plans

Man-made and topographic (natural) features and other relevant project information, including the information needed to correctly locate the structure on the site, are shown on a site plan (commonly called a plot plan). Man-made features include roads, sidewalks, utilities, and buildings. Topographical features include trees, streams, springs, and existing contours. Project information includes the building outline, general utility information, proposed sidewalks, parking areas, roads, landscape information, proposed contours, and any other information that will convey what is to be constructed or changed on the site. A prominently displayed north direction arrow is included for orientation purposes on site plans. Sometimes a site plan contains a large-scale map of the overall area that indicates where the project is located on the site. *Figures 3* and *4* show examples of basic site plans.

Typically, site plans show the following types of detailed information:

- Coordinates of control points or property corners
- Direction and length of **property lines** or control lines

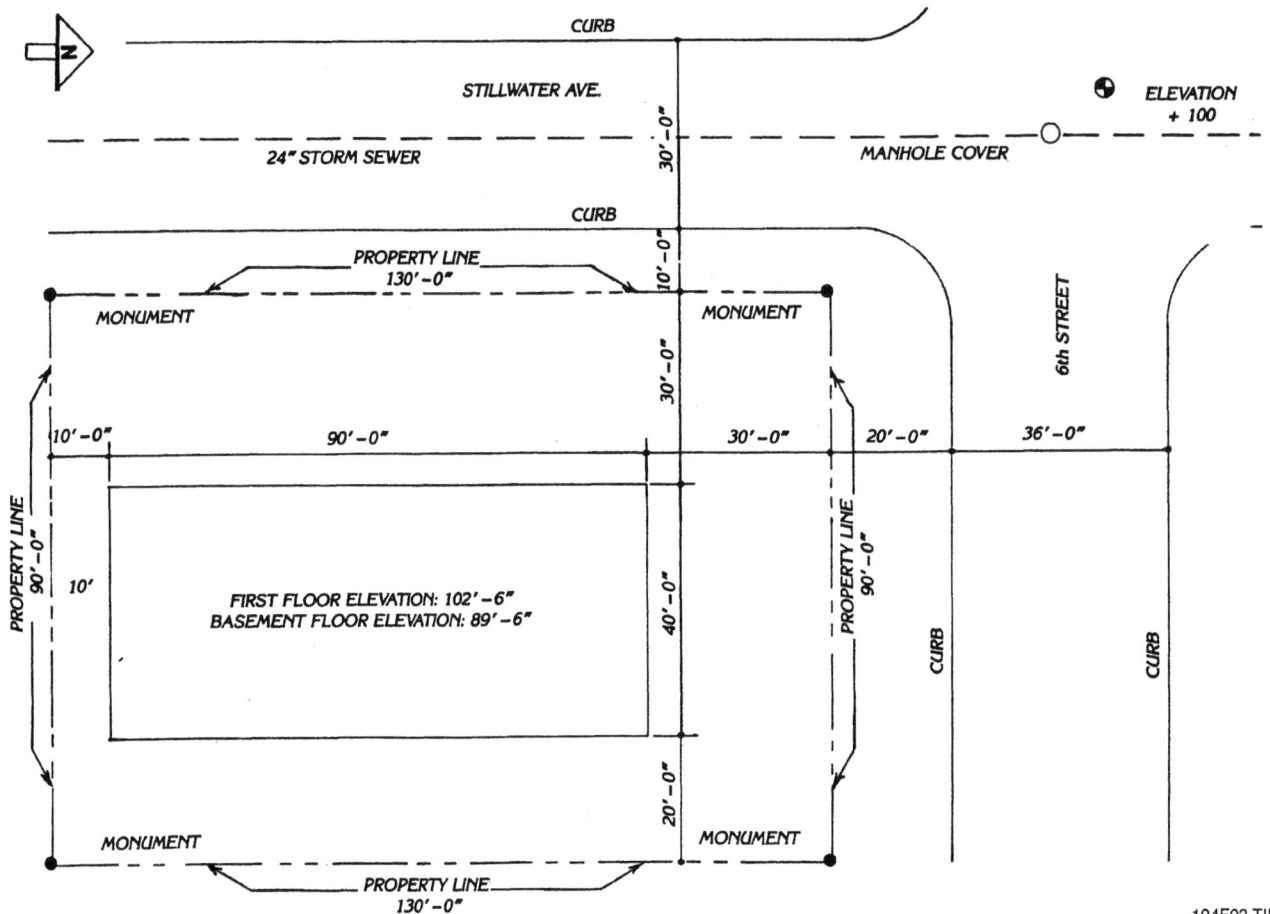

Figure 3 ◆ Site plan.

104F03.TIF

- Description, or reference to a description, for all control and property **monuments**
- Location, dimensions, and elevation of the structure on site
- Finish and existing grade **contour lines**
- Location of utilities
- Location of existing elements such as trees and other structures
- Location and dimensions of roadways, driveways, and sidewalks
- Names of all roads shown on the plan
- Locations and dimensions of any **easements**

Like other drawings, site plans are drawn to scale. The scale used depends on the size of the project. A project covering a large area typically will have a small scale, such as 1" = 100', while a project on a small site might have a large scale, such as 1" = 10'.

Normally, the dimensions shown on site plans are stated in feet and tenths of a foot (engineer's scale). However, some site plans state the dimensions in feet, inches, and fractions of an inch (architect's scale). Dimensions to the property lines are

shown to establish code requirements. Frequently, building codes require that nothing be built on certain portions of the land. For example, local building codes may have a **front setback** requirement that dictates the minimum distance that must be maintained between the street and the front of a structure. Normally, side yards have a minimum width specified from the property line to allow for access to rear yards and to reduce the possibility of fire spreading to adjacent buildings.

A property owner cannot build on an area where an easement has been identified. Examples of typical easements are the right of a neighbor to build a road; a public utility to install water, gas, or electric lines on the property; or an area set aside for drainage of groundwater.

Site plans show finish grades (also called elevations) for the site based on data provided by a surveyor or engineer. It is necessary to know these elevations for grading the lot and for construction of the structure. Finish grades are typically shown for all four corners of the lot as well as other points within the lot. Finished grades or elevations are also shown for the corners of the structure and relevant points within the building.

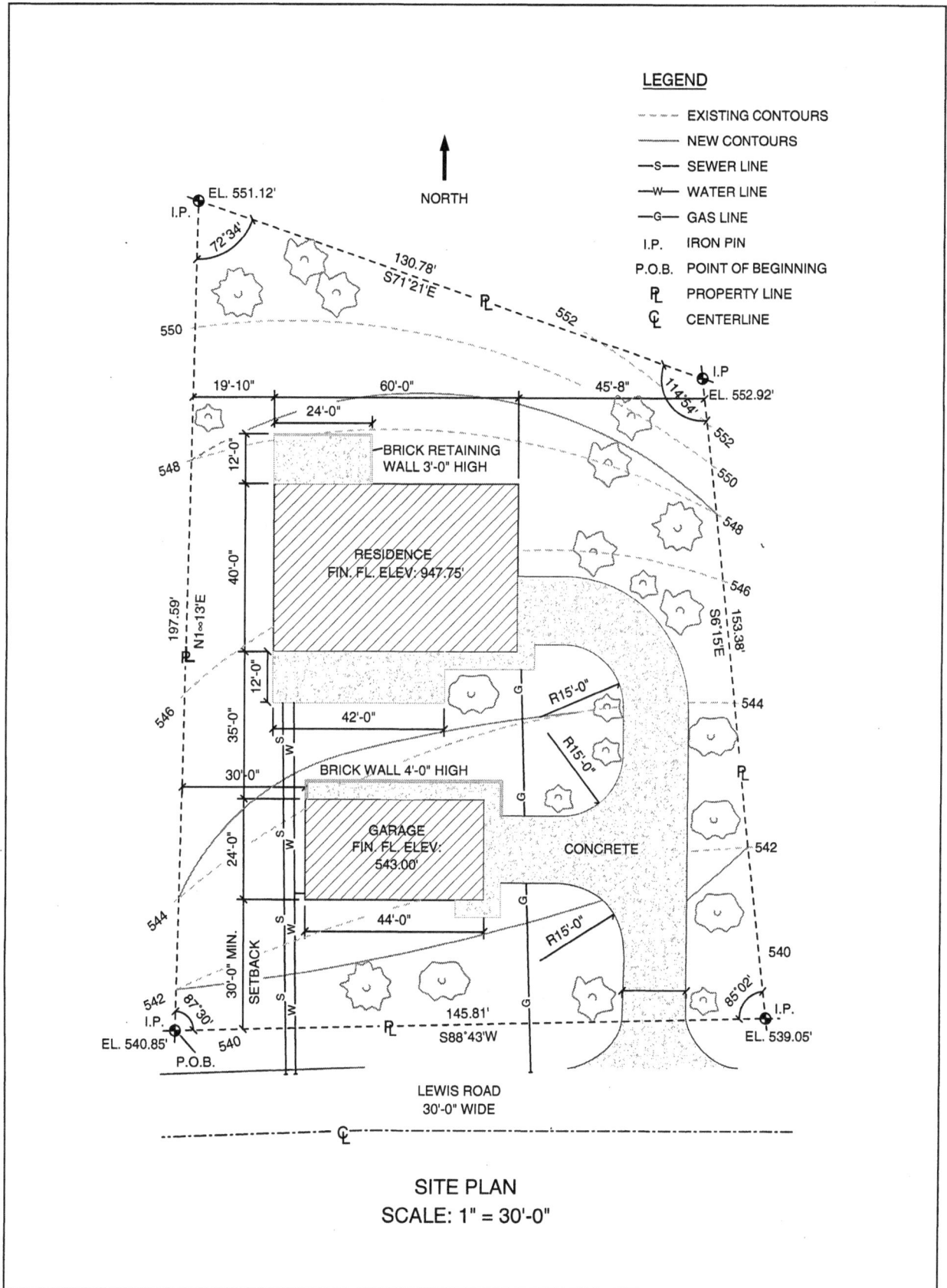

LEGEND

- – – – – – EXISTING CONTOURS
- ———— NEW CONTOURS
- —S— SEWER LINE
- —W— WATER LINE
- —G— GAS LINE
- I.P. IRON PIN
- P.O.B. POINT OF BEGINNING
- P⎣ PROPERTY LINE
- ℄ CENTERLINE

EL. 551.12'
I.P.
72°34'
130.78'
S71°21'E
P⎣
552
550

NORTH

I.P
EL. 552.92'
114°54'
552
550
548
546
S61°15'E
153.38'

19'-10"   60'-0"   45'-8"
24'-0"
12'-0"

BRICK RETAINING
WALL 3'-0" HIGH

548

40'-0"

RESIDENCE
FIN. FL. ELEV: 947.75'

546

544

12'-0"

42'-0"

R15'-0"

CONCRETE

197.59'
N1°13'E

35'-0"

30'-0"

BRICK WALL 4'-0" HIGH

R15'-0"

546

544

24'-0"

GARAGE
FIN. FL. ELEV:
543.00'

P⎣

542

44'-0"

R15'-0"

540

30'-0" MIN.

SETBACK

542
I.P.
87°30'

85°02'
I.P.
EL. 539.05'

EL. 540.85'

540

145.81'
P⎣
S88°43'W

P.O.B.

LEWIS ROAD
30'-0" WIDE

℄

SITE PLAN
SCALE: 1" = 30'-0"

104F04.EPS

Figure 4 ◆ Site plan showing topographical features.

All the finish grade references shown are keyed to a reference point, called a bench mark or job datum. This is a reference point established by the surveyor on or close to the property, usually at one corner of the lot. At the site, this point may be marked by a plugged pipe driven into the ground, a brass marker, or a wood stake. The location of the bench mark is shown on the plot plan with a grade figure next to it. This grade figure may indicate the actual elevation relative to sea level, or it may be an arbitrary elevation, such as 100.00' or 500.00'. All other grade points shown on the site plan, therefore, will be relative to the bench mark.

A site plan usually shows the finish floor elevation of the building. This is the level of the first floor of the building relative to the job-site bench mark. For example, if the bench mark is labeled 100.00' and the finish floor elevation indicated on the plan is marked 105.00', the finished floor elevation is 5' above the bench mark. During construction, many important measurements are taken from the finish floor elevation point.

2.2.2 Foundation Plans

As applicable, foundation plans (*Figure 5*) give information about the location and dimensions of footings, grade beams, foundation walls, stem walls, piers, equipment footings, and windows and doors. The specific information shown on the plan is determined by the type of construction involved: full-basement foundation, crawl space, or a concrete slab-on-grade level (*Figure 6*).

The following are types of information normally shown on foundation plans for full-basement and crawl space foundations:

- Location of the inside and outside of the foundation walls
- Location of the footings for foundation walls, columns, posts, chimneys, fireplaces, etc.
- Wall openings for windows, doors, crawl space access, and vents
- Walls for entrance platforms (stoops)
- Floor joist direction, size, and spacing

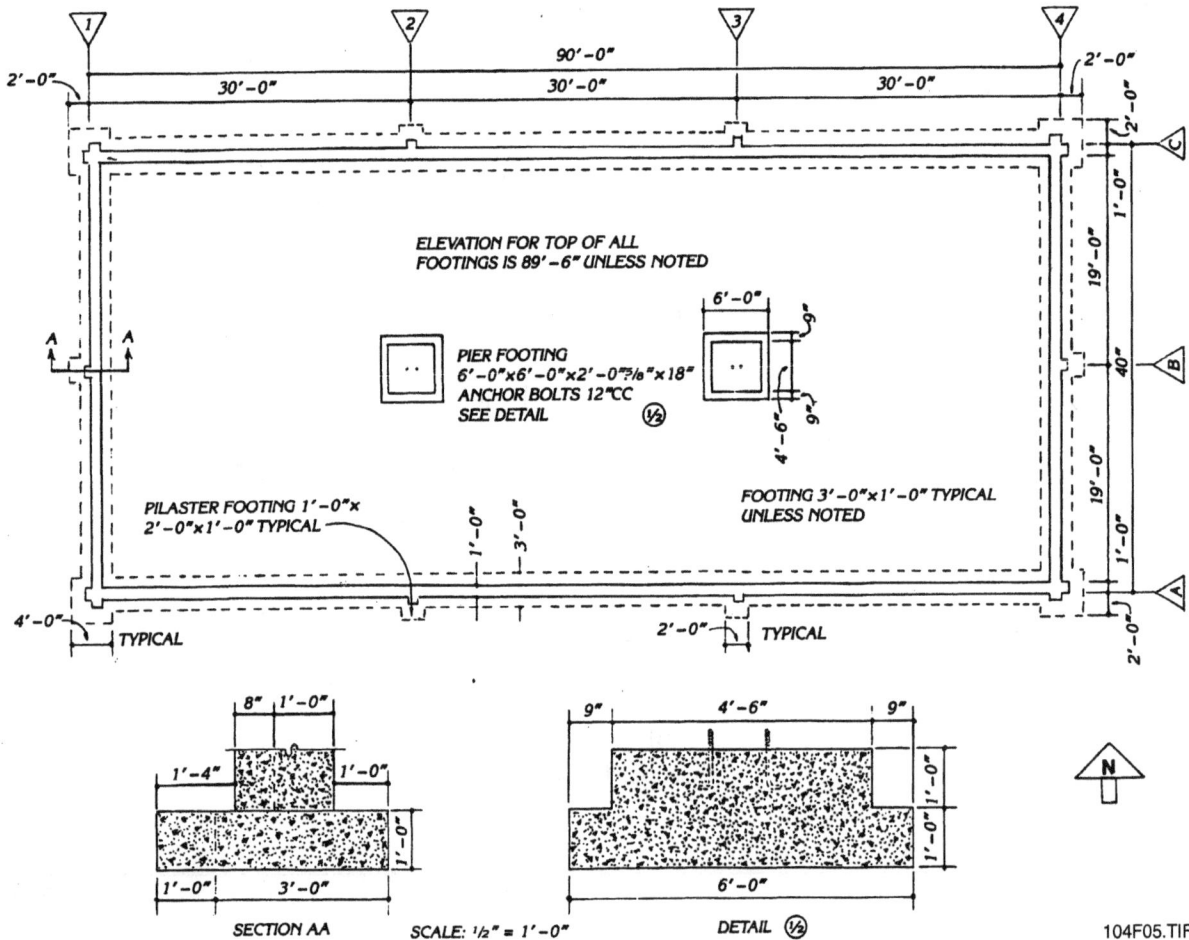

Figure 5 ◆ Foundation plan.

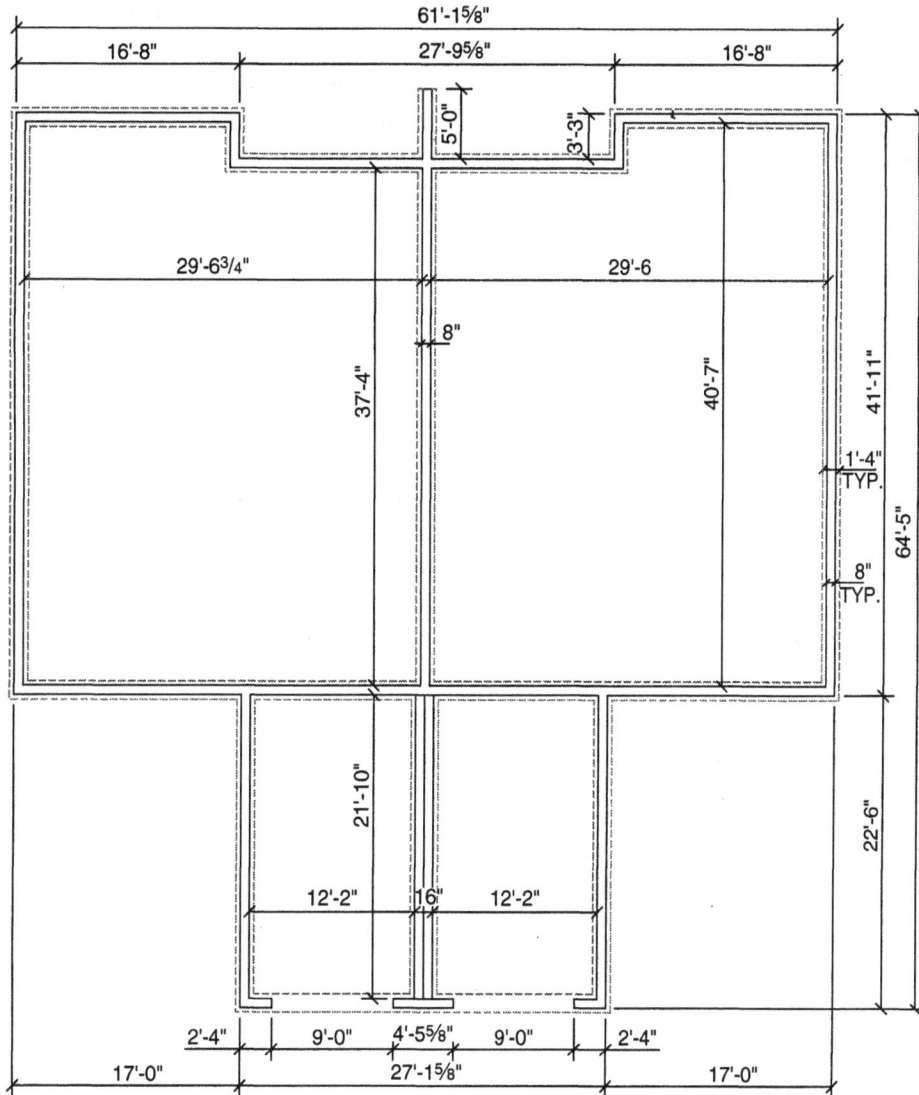

FOUNDATION PLAN
Scale: ³⁄₃₂" = 1'-0"

104F06.EPS

Figure 6 ◆ Slab-on-grade plan.

- Locations of stairways
- Notations for the strength of concrete used for various parts of the foundation and floor
- Notations for the composition, thickness, and underlaying material of the basement floor or crawl space surface
- Location of furnaces and other equipment

The types of information normally shown on foundation plans for slab-on-grade foundations include the following:

- Size and shape of the slab
- Exterior and interior footing locations
- Location of fireplaces, floor drains, heating, ventilating, and air conditioning (HVAC) ductwork, etc.

- Notations for slab thickness
- Notations for wire mesh reinforcing, fill, and vapor barrier materials

2.2.3 Floor Plans

The floor plan is the main drawing of the entire set. For a floor plan view, an imaginary line is cut horizontally across the structure at varying heights so all the important features such as windows, doors, and plumbing fixtures can be shown. For multistory buildings, separate floor plans are normally drawn for each floor. However, if several floors have the same layout, one drawing may be used to show all the floors that are similar. *Figure 7* shows an example of a basic floor plan. The types of information normally shown on floor plans include the following:

- Outside walls, including the locations and dimensions of all exterior openings
- Types of construction materials
- Locations of interior walls and partitions
- Locations and swing of doors
- Stairways
- Locations of windows
- Locations of cabinets, electrical and mechanical equipment, and fixtures
- Location of cutting plane line

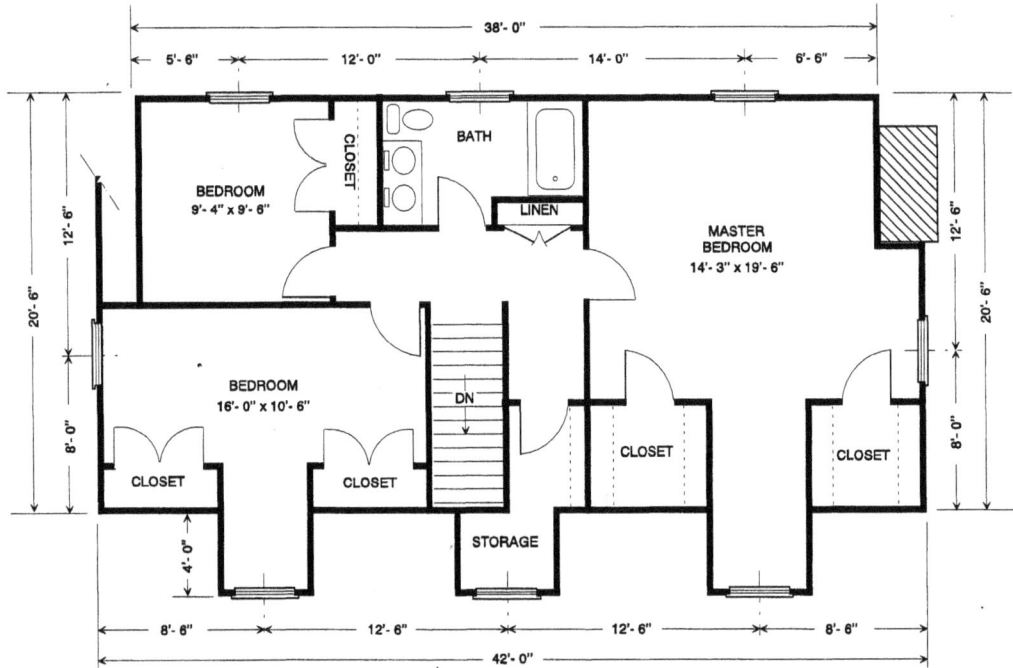

UPPER LEVEL

FLOOR PLAN

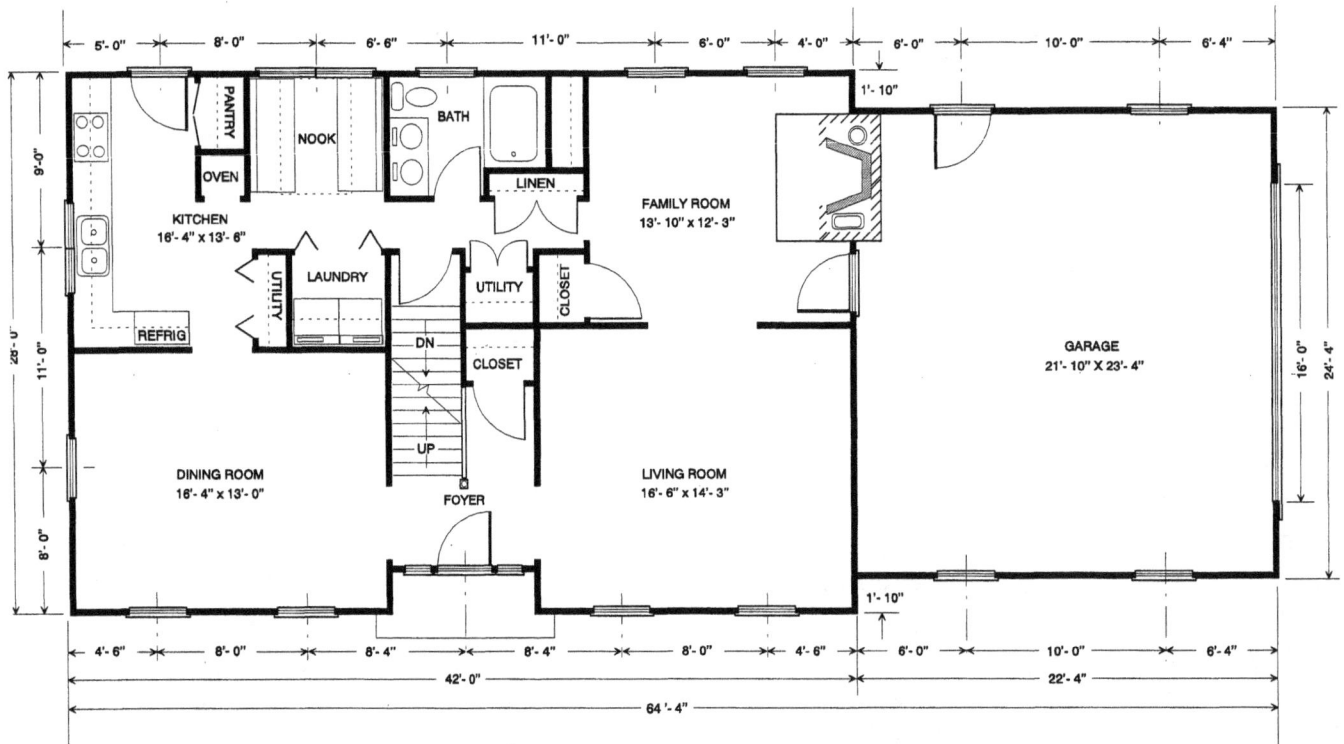

LOWER LEVEL

104F07.EPS

Figure 7 ◆ Floor plans for a building.

Each door or window shown on a floor plan for a commercial building is typically accompanied by a number, letter, or both. This number/letter is an identifier that refers to a door or window schedule that describes the corresponding size, type of material, model number, etc., for the specific door or window. Door and window schedules are discussed in more detail later in this section. Residential floor plans often show door sizes directly on the plan drawing.

2.2.4 Roof Plans

When supplied, roof plans (*Figure 8*) provide information about the roof slope, roof drain placement, and other pertinent information regarding ornamental sheet metal work, gutters and downspouts, etc. Where applicable, the roof plan may also show information on the location of air conditioning units, exhaust fans, and other ventilation equipment.

Some drawing sets may also contain a ceiling plan that shows the locations of supply diffusers, exhaust grilles, access panels, and the locations of structural components and other mechanicals.

Some have reflected ceiling plans. These show the details of the ceiling as though it were reflected by a mirror on the floor. Reflected ceiling plans show features of the ceiling while keeping those features in proper relation to the floor plan. For example, if a vertical pipe runs from floor to ceiling in a room and is drawn in the upper left corner of the floor plan, it is also shown on the upper left corner of the reflected ceiling plan of that same room.

2.3.0 Elevation Drawings

Elevation drawings are views that look straight ahead at a structure. The object is projected from a vertical plane. Typically, **elevation views** are used to show the exterior features of a structure so that the general size and shape of the structure can be determined. Elevation drawings clarify much of the information on the floor plan. For example, a floor plan shows where the doors and windows are located in the outside walls; an elevation view of the same wall shows actual representations of these doors and windows. *Figure 9* shows an example of a basic elevation drawing.

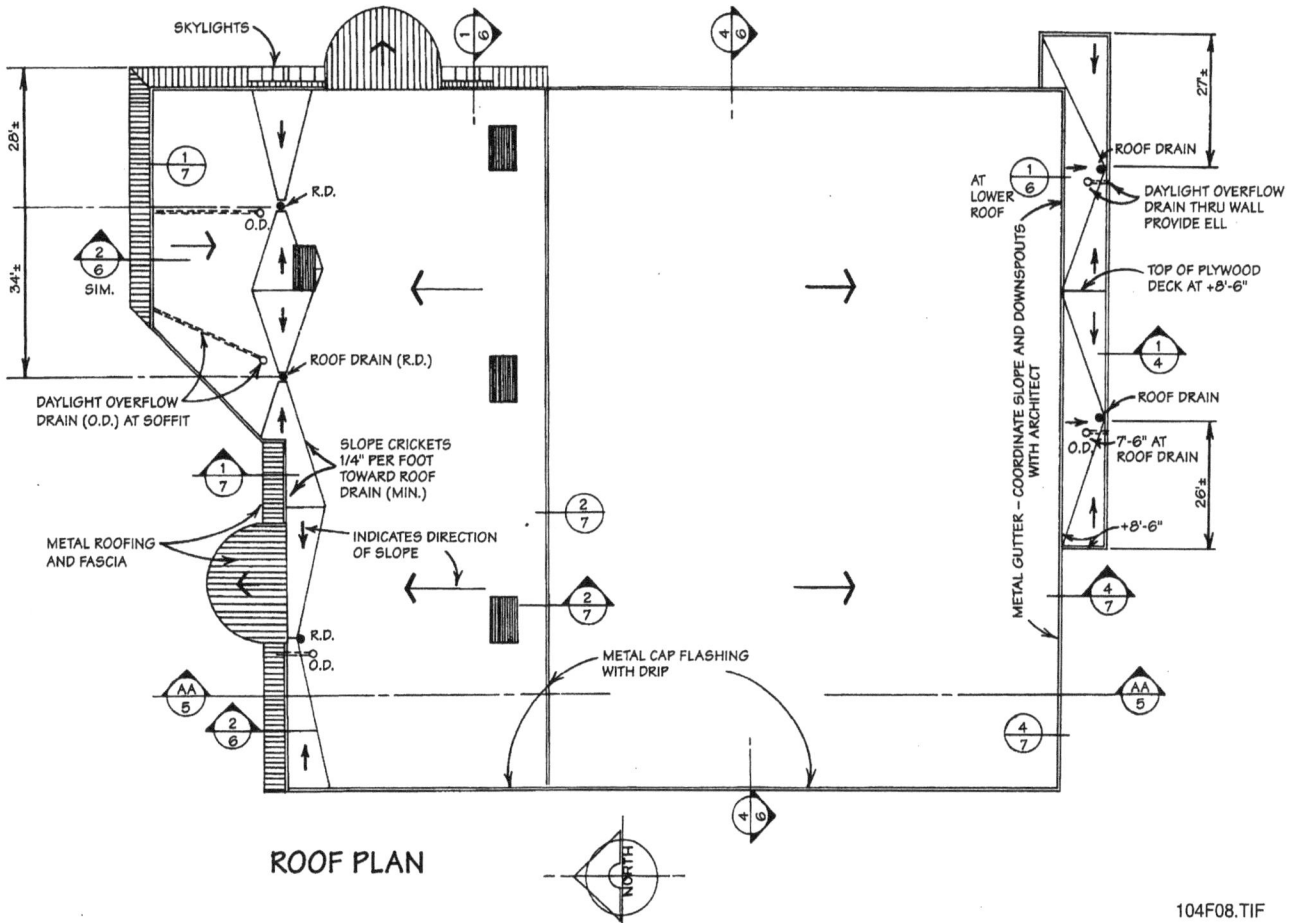

Figure 8 ◆ Roof plan.

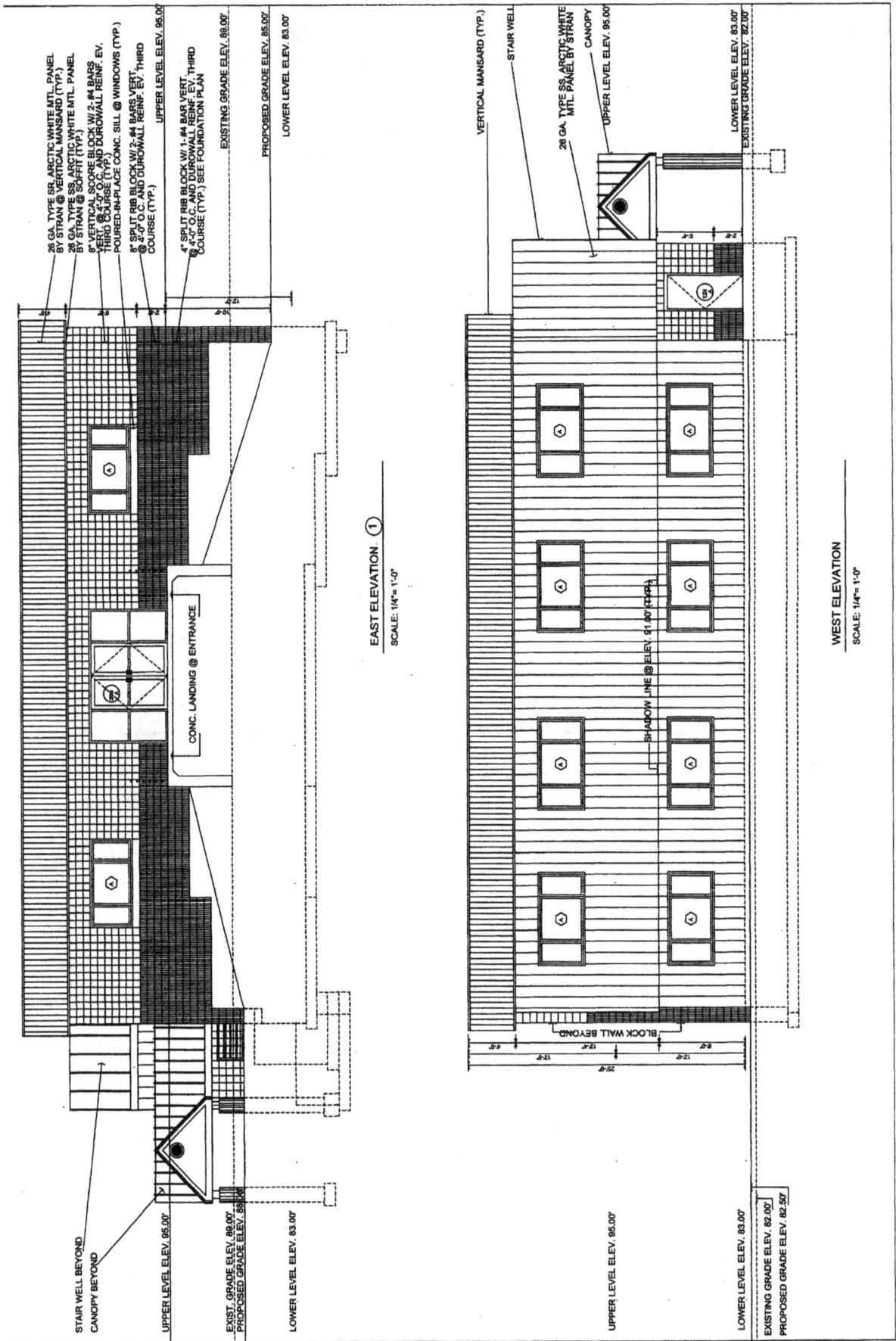

Figure 9 ◆ Elevation drawing.

104F09.TIF

The following types of information are normally shown on elevation drawings:

- Grade lines
- Floor height
- Window and door types
- Roof lines and slope, roofing material, vents, gravel stops, and projection of eaves
- Exterior finish materials and trim
- Exterior dimensions

Unless one or more views are identical, four elevation views are generally used to show each exposure. With very complex buildings, more than four views may be required. The various views are usually labeled in one of two ways. They may be labeled as front, right side, left side, and rear, or they may be designated by compass direction. For example, if the front of the building faces north, then this becomes the north elevation. The other elevations are then labeled accordingly (east, south, and west).

Drawing sets may also have interior elevation views for the walls in each partitioned area, especially for walls that have special features.

2.4.0 Section Drawings

A section view or drawing (*Figure 10*) shows how a particular feature looks inside or how something is put together internally. The feature is drawn as if a cut has been made through the middle of it at a certain point. The location of the cut and the direction to be viewed are shown on the related plan view.

Section drawings that show a view made by cutting through the length of a structure are referred to as longitudinal sections, while those showing the view of a cut through the width or narrow portion of the structure are referred to as traverse sections. To show greater detail, section views are normally drawn to a larger scale than that used in plan views. The following types of information are normally shown by a section view:

- Details of construction and information about stairs, walls, chimneys, or other parts of construction that may not show clearly on a plan view
- Floor levels in relation to grade
- Wall thickness at various locations
- Anchors and reinforcing steel

2.5.0 Detail Drawings

Details are enlargements of special features of a building or of equipment installed in a building. They are drawn to a larger scale in order to make the details clearer. The detail drawings are often placed on the same sheet where the feature appears in the plan. *Figure 11* shows a typical detail view.

Typically, details may be drawn for the following objects or situations:

- Footings and foundations, including anchor bolts, reinforcing, and control joints
- Beams, floor joists, bridging, and other support members
- Sills, floor framing, exterior walls, and vapor barriers
- Floor heights, thickness, expansion, and reinforcing
- Interior walls
- Windows, exterior and interior doors, and door frames
- Roofs, cornices, soffits, and ceilings
- Gravel stops, fascia, and flashing
- Fireplaces and chimneys
- Staircases and stair assemblies
- Millwork, trim, ornamental iron, and specialty items

2.6.0 Schedules

Schedules in a drawing set are tables that describe and specify the various types and sizes of materials used in the construction of a building. Commonly, there are finish schedules (*Figure 12*), as well as door and window schedules (*Figure 13*).

In the finish schedule for a structure, each room is identified by name or number. The material and finish for each part of the room (walls, floor, ceiling, base, and trim) are designated, along with any clarifying remarks.

Door and window types are identified on the various plan and elevation drawings by numbers, letters, or both. The door and window schedules list these identifier numbers or letters and describe the corresponding size, type of material, and model number for each different type of door or window used in the structure.

For commercial jobs, additional schedules are provided for mechanical equipment and controls, plumbing fixtures, lighting fixtures, and any other equipment that needs to be listed separately. *Figure 14* shows schedules typical of those found on mechanical plans.

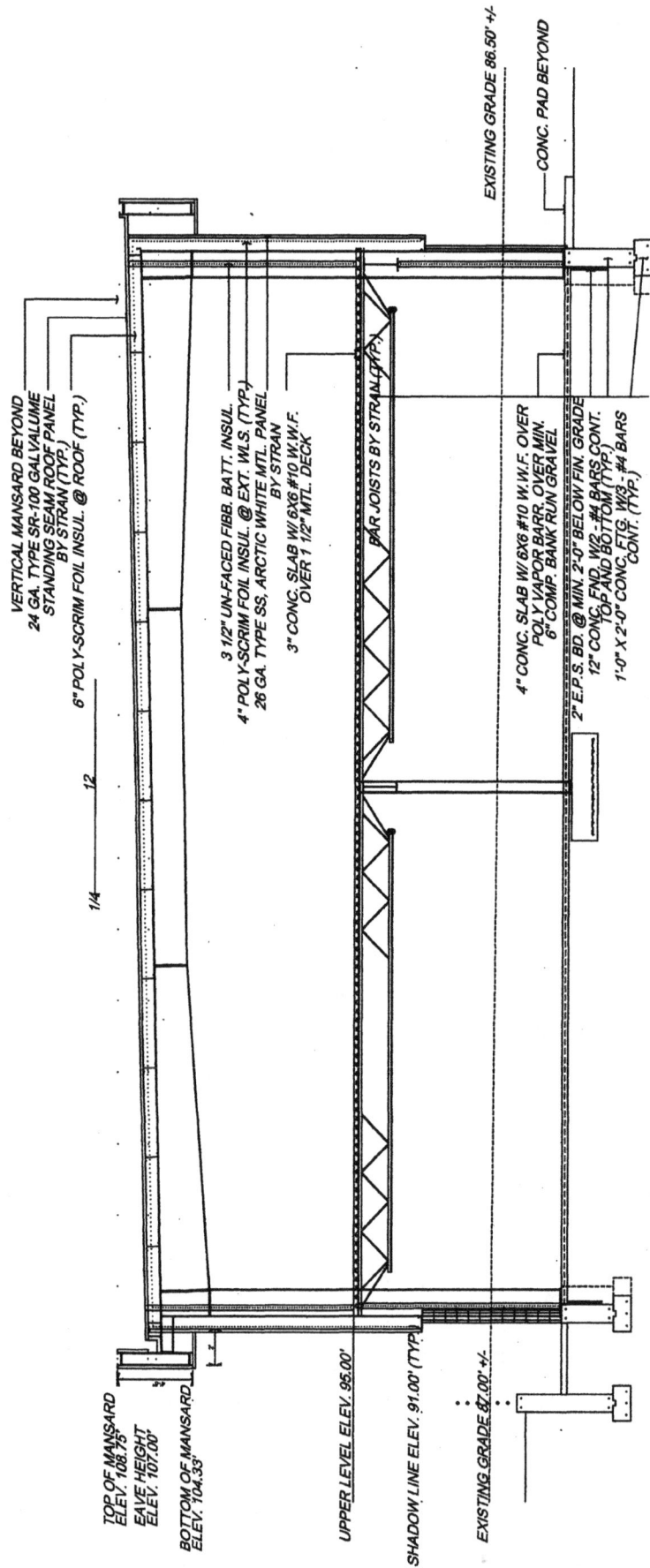

VERTICAL MANSARD BEYOND
24 GA. TYPE SR-100 GALVALUME
STANDING SEAM ROOF PANEL
BY STRAN (TYP.)
6" POLY-SCRIM FOIL INSUL. @ ROOF (TYP.)

3 1/2" UN-FACED FIBB. BATT. INSUL.
4" POLY-SCRIM FOIL INSUL. @ EXT. WLS. (TYP.)
26 GA. TYPE SS, ARCTIC WHITE MTL. PANEL
BY STRAN
3" CONC. SLAB W/ 6X6 #10 W.W.F.
OVER 1 1/2" MTL. DECK

BAR JOISTS BY STRAN (TYP.)

4" CONC. SLAB W/ 6X6 #10 W.W.F. OVER
POLY VAPOR BARR. OVER MIN.
6" COMP. BANK RUN GRAVEL
2" E.P.S. BD. @ MIN. 2'-0" BELOW FIN. GRADE
12" CONC. FND. W/3 - #4 BARS CONT.
TOP AND BOTTOM (TYP.)
1'-0" X 2'-0" CONC. FTG. W/3 - #4 BARS
CONT. (TYP.)

EXISTING GRADE 86.50' +/-
CONC. PAD BEYOND

12
1/4

TOP OF MANSARD
ELEV. 108.75'
EAVE HEIGHT
ELEV. 107.00'
BOTTOM OF MANSARD
ELEV. 104.33'

UPPER LEVEL ELEV. 95.00'

SHADOW LINE ELEV. 91.00' (TYP.)

EXISTING GRADE 87.00' +/-

BUILDING SECTION
SCALE: 1/4" = 1'-0"

104F10.TIF

Figure 10 ◆ Section drawing showing building construction.

4.12

SITE LAYOUT LEVEL ONE — TRAINEE MODULE 78104-04

PROVIDE 5" BRG.
DEPTH AT K SERIES
JOISTS

1½" ROOF DECK

⅛" ⎰ 2"

T.O. CONC.
EL. SEE PLAN

11"

T.O. STEEL
EL. SEE PLAN

⁵⁄₁₆ φ ROD FOR FIRST
(2) JOISTS EACH SIDE
OF COLUMN (8 TOTAL
PER COLUMN)

³⁄₁₆" ⎰ 1"

28K
RE: PLAN

WF
RE: PLAN

TS COLUMN
RE: PLAN

LH
RE: PLAN

⅜" STIFFENER PLATE
EACH SIDE

½" CAP ℞ W/
(4)—¾"φ BOLTS

SECTION ③③/S3 ¾"=1'-0

104F11.EPS

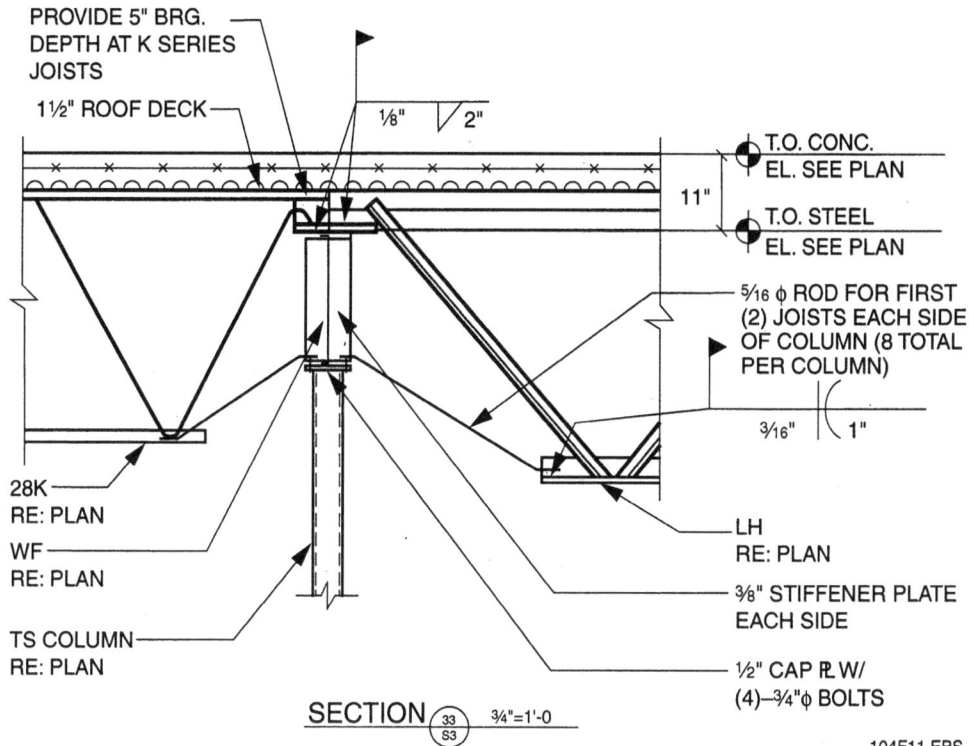

Figure 11 ◆ Detail drawing.

ROOM FINISH SCHEDULE																	
ROOMS	FLOOR				CEILING				WALL				BASE			TRIM			REMARKS
	CARPET	CERAMIC TILE	RUBBER TILE	CONCRETE	ACOUSTIC TILE	DRYWALL	PAINT	CERAMIC TILE	DRYWALL	PAINT	WALLPAPER	CERAMIC TILE	WOOD	RUBBER	CERAMIC TILE	STAIN	WOOD	STAIN	PAINT	
ENTRY		✓			✓				✓	✓	✓		✓			✓	✓	✓		See owner for all painting
HALL	✓				✓				✓	✓			✓			✓	✓	✓		
BEDROOM 1	✓				✓				✓	✓	✓		✓			✓	✓	✓		See owner for grade of carpet
BEDROOM 2	✓				✓				✓	✓			✓			✓	✓	✓		See owner for grade of carpet
BEDROOM 3	✓				✓				✓	✓			✓			✓	✓	✓		See owner for grade of carpet
BATH 1	✓	✓			✓			✓	✓	✓	✓	✓				✓	✓	✓		Wallpaper 3 walls around vanity
BATH 2		✓			✓			✓	✓	✓	✓	✓		✓			✓	✓		Water-seal tile Wallpaper w/wall
UTIL + CLOSETS	✓		✓			✓	✓		✓	✓			✓			✓	✓	✓	✓	Use off-white flat latex
KITCHEN		✓			✓				✓	✓		✓					✓	✓		
DINING	✓				✓				✓	✓	✓		✓			✓	✓	✓		
LIVING	✓				✓				✓	✓			✓			✓	✓	✓		See owner for grade of carpet
GARAGE				✓		✓	✓				✓		✓				✓	✓		

104F12.EPS

Figure 12 ◆ Example of a finish schedule.

DOOR SCHEDULE

DOOR	WIDTH	HEIGHT	THICK-NESS	MAT'L	TYPE	STORM DOOR	QTY.	THRES-HOLD	REMARKS	MANUFACTURER
2068	2'-0"	6'-8"	1 3/8"	Wood-Ash	Hollow-core	NO	5	None	Oil Stain	LBJ Door Co.
2468	2'-4"	6'-8"	1 3/8"	Wood-Ash	Hollow-core	NO	1	None	Oil Stain	LBJ Door Co.
2668	2'-6"	6'-8"	1"	Wood-Ash	Cafe	NO	1 pr.	None	Oil Stain	LBJ Door Co.
2668	2'-6"	6'-8"	1 3/8"	Wood-Ash	Sliding Pocket	NO	1	None	Oil Stain	LBJ Door Co.
2668	2'-6"	6'-8"	1 3/8"	Wood-Ash	Hollow-core	1 Screen	6	Alum.	Screen door in garage	LBJ Door Co.
2868	2'-8"	6'-8"	1 3/4"	Metal Clad	Fireproof	YES	1	Alum.	Paint	LBJ Door Co.
3068	3'-0"	6'-8"	1 3/4"	Wood-Ash	Solid-core	NO	1	None	Oil Stain	LBJ Door Co.
3668	3'-6"	6'-8"	1 3/4"	Wood-Ash	Solid-core	YES	1	Alum.	Marine Varnish	LBJ Door Co.
6066	6'-0"	6'-6"	1/2"	Glass/Metal	Sliding	YES	1 pr.	Alum.	Sliding Screen	LBJ Door Co.
6068	6'-0"	6'-8"	1 1/4"	Wood-Ash	Bi-Fold	NO	2 sets	None	Oil Stain	LBJ Door Co.
1956	1'-9"	5'-6"	1/2"	Glass/Metal	Sliding Shower door	NO	2 sets	None	Frosted Glass	LBJ Door Co.

WINDOW SCHEDULE

SYMBOL	WIDTH	HEIGHT	MAT'L	TYPE	SCREEN & STORM	QUANTITY	REMARKS	MANUFACTURER	CATALOG NUMBER
A	3'-8"	3'-0"	ALUM.	DOUBLE HUNG	YES	2	4 LIGHTS, 4 HIGH	LBJ Window Co.	141 PW
B	3'-8"	5'-0"	ALUM.	DOUBLE HUNG	YES	1	4 LIGHTS, 4 HIGH	LBJ Window Co.	145 PW
C	3'-0"	5'-0"	ALUM.	STATIONARY	STORM ONLY	2	SINGLE LIGHTS	H & J Glass Co.	59 PY
D	2'-0"	3'-0"	ALUM.	DOUBLE HUNG	YES	1	4 LIGHTS, 4 HIGH	LBJ Window Co.	142 PW
E	2'-0"	6'-0"	ALUM.	STATIONARY	STORM ONLY	2	20 LIGHTS	H & J Glass Co.	37 TS
F	3'-6"	5'-0"	ALUM.	DOUBLE HUNG	YES	1	16 LIGHTS, 4 HIGH	LBJ Window Co.	143 PW

HEADER SCHEDULE

HEADER SIZE	EXTERIOR		INTERIOR	
	26' + UNDER	26' TO 32'	26' + UNDER	26' TO 32'
(2) 2 x 4	3'-6"	3'-0"	USE (2) 2 x 6	
(2) 2 x 6	6'-6"	6'-0"	4'-0"	3'-0"
(2) 2 x 8	8'-6"	8'-0"	5'-6"	5'-0"
(2) 2 x 10	11'-0"	10'-0"	7'-0"	6'-6"
(2) 2 x 12	13'-6"	12'-0"	8'-6"	8'-0"

104F13.EPS

Figure 13 ◆ Examples of door, window, and header schedules.

2.7.0 Structural Drawings

Structural drawings are created by a structural engineer and accompany the architect's plans. They are usually drawn for large structures such as office buildings or factories. They show requirements for structural elements of the building, including columns, floor and roof systems, stairs, canopies, and bearing walls.

Structural drawings contain details such as:

- Heights of finished floors and walls
- Height and bearing of bar joist or steel joist
- Locations of bearing steel materials
- Height of steel beams, concrete plank, concrete Ts, and poured-in-place concrete
- Bearing plate locations

CABINET UNIT HEATER SCHEDULE

UNIT HEATER NO.	LOCATION	C.F.M.	FAN MOTOR				MBH	GPM	EWT	EAT	MAX. WATER P.D.	REMARKS
			H.P.	VOLTS	PHASE	Hz						
CUH-1	124	400	$^1/_{12}$	115	1	60	23	2.3	180°F	60°F	2.7	McQUAY #CHF004 SEMI-RECESSED, R.H COIL
CUH-2	137	400	$^1/_{12}$	115	1	60	23	2.3	180°F	60°F	2.7	McQUAY #CHF004 SEMI-RECESSED, L.H COIL
CUH-3	143	400	$^1/_{12}$	115	1	60	23	2.3	180°F	60°F	2.7	McQUAY #CHF004 SEMI-RECESSED, R.H COIL

NOTES:
1. 3 Speed Control
2. Front Discharge
3. With Return Air Filters

PUMP SCHEDULE

UNIT NO.	LOCATION	SERVICE.	GPM	MBH	MOTOR					TYPE	REMARKS
					RPM	H.P.	VOLTS	PHASE	Hz		
P-1	MECH. ROOM	NAVE	45	41'	1750	$1^1/_2$	208/230	3	60	IN-LINE	B/G #60-20T SERVICE 40% GLYCOL SOLUTION
P-2	MECH. ROOM	CHW TO AHU 1-4	65	37'	1750	$1^1/_2$	208/230	3	60	IN-LINE	B/G #60-20T SERVICE 40% GLYCOL SOLUTION
P-3	MECH. ROOM	RECIR. TO TANK	40	17'	1750	$^1/_2$	208/230	3	60	IN-LINE	B/G #60-13T SERVICE 40% GLYCOL SOLUTION
P-4	EXIST. MECH. ROOM	HW	73	31'	1750	$1^1/_2$	208/230	3	60	IN-LINE	B/G #60-20T HOT WATER

NOTES:
1. Starters And Disconnects By E.C.

GRILLE, REGISTER AND DIFFUSER SCHEDULE

ITEM	MANUFACTURER	MODEL NO.	QTY.	LOCATION	CFM EACH	AIR PATTERN	SIZE		FINISHES	REMARKS
							FRAME	NECK		
A	BARBER COLMAN	SFSV	8	126,127,128 144	245	4-WAY	12"× 12"	8"Ø	#7 OFF-WHITE	
B	BARBER COLMAN	SFSV	2	142	275	4-WAY	18"× 18"	10"Ø	#7 OFF-WHITE	
C	BARBER COLMAN	SFSV	4	140, 141	240	4-WAY	12"× 12"	8"Ø	#7 OFF-WHITE	
D	BARBER COLMAN	SFSV	2	139	270	4-WAY	18"× 18"	10"Ø	#7 OFF-WHITE	
E	BARBER COLMAN	SFSV	2	138	280	4-WAY	18"× 18"	10"Ø	#7 OFF-WHITE	
F	BARBER COLMAN	SFSV	2	136	250	4-WAY	12"× 12"	6"Ø	#7 OFF-WHITE	
G	BARBER COLMAN	SFSV	2	135	235	4-WAY	12"× 12"	6"Ø	#7 OFF-WHITE	
H	BARBER COLMAN	SFSV	3	134	100	4-WAY	12"× 12"	6"Ø	#7 OFF-WHITE	FIRE DAMPER SEE DETAIL A
I	BARBER COLMAN	SFSV	1	134	190	4-WAY	12"× 12"	8"Ø	#7 OFF-WHITE	FIRE DAMPER SEE DETAIL A
		SFSV	1	134			12"×			FIRE DAMPER

104F14.EPS

Figure 14 ◆ Mechanical equipment schedules.

- Location, size, and spacing of anchor bolts
- Stairways

2.8.0 Plumbing Plans

Plumbing plans show the layout of fixtures, water supply lines, and lines to sewage disposal systems. The plans may be included in the floor plan of a regular construction job or on a separate plan for a large commercial structure. When drawn as a separate plan, the plumbing plan details are usually overlaid on tracings of the various building floor plans from which unnecessary details have been omitted to allow the location and layout of the plumbing systems to show clearly. *Figure 15* is a plumbing plan showing the sanitary plumbing.

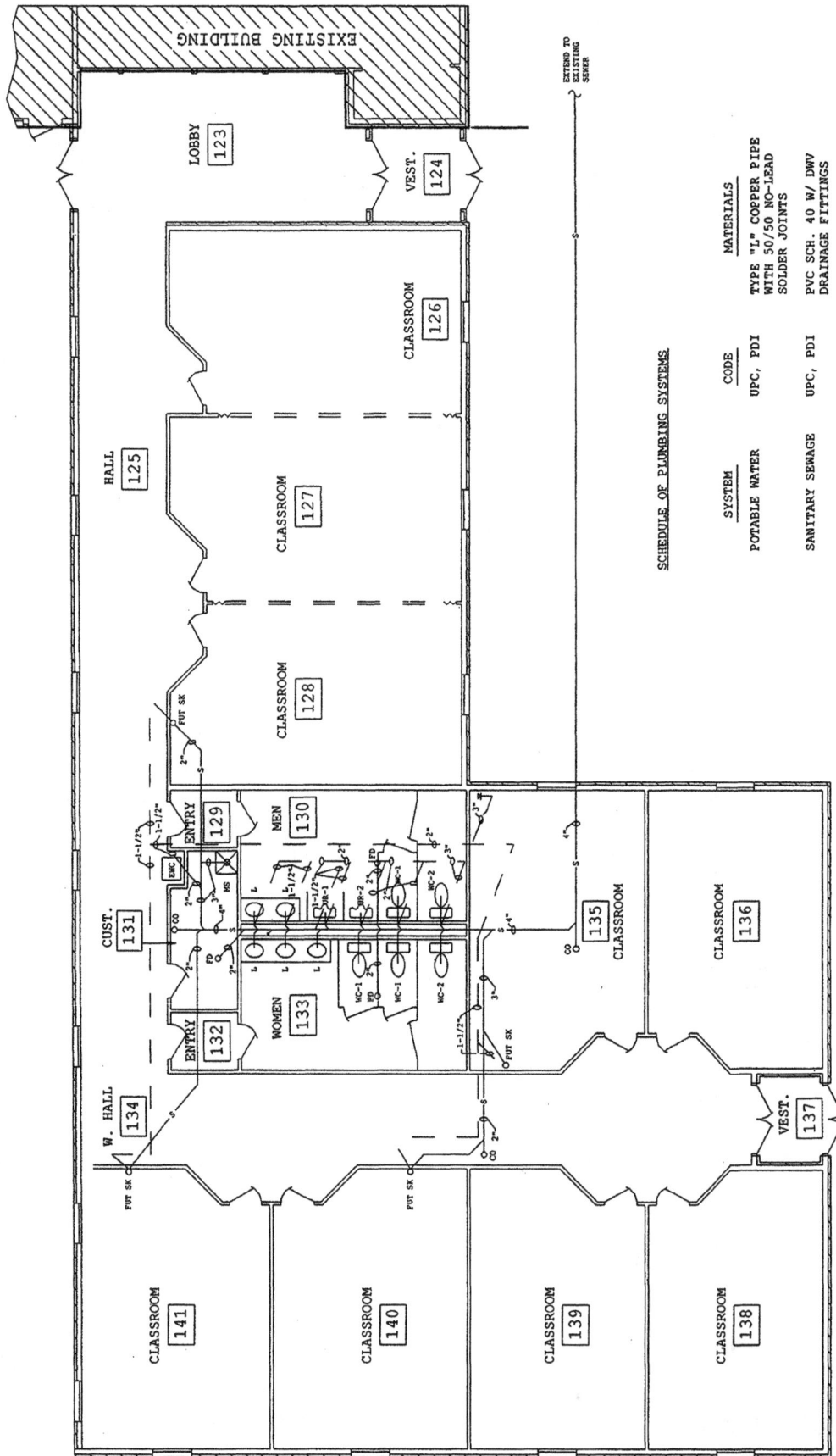

Figure 15 ◆ Sanitary plumbing plan. (1 of 2)

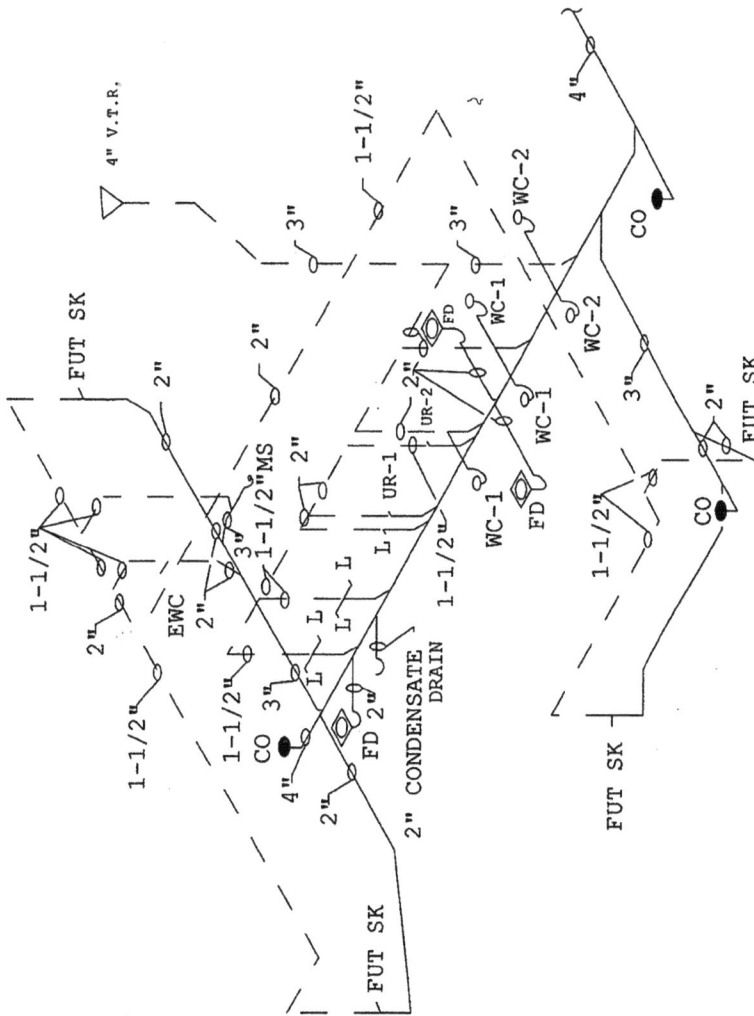

PLUMBING SYSTEM SPECIFICATIONS

COMPLY WITH APPLICABLE STATE AND LOCAL PLUMBING CODES AND STANDARDS
PERTAINING TO MATERIALS, PRODUCTS AND INSTALLATION OF POTABLE WATER
AND SANITARY SEWAGE SYSTEMS.

TEST EACH PLUMBING SYSTEM IN ACCORDANCE WITH APPLICABLE CODES AND
STANDARDS, STERILIZE POTABLE WATER SYSTEMS PER STATE AND LOCAL
UTILITY REQUIREMENTS.

SANITARY/VENT ISOMETRIC

104F15B.TIF

Figure 15 ◆ Sanitary plumbing plan. (2 of 2)

Many plumbing plans show a separate **riser diagram** (*Figure 16*) that is often not drawn to scale. A plumbing legend shows the various symbols pertaining to the plan. Common plumbing symbols are covered later in this module. Some legends provide tabulated plumbing fixture and equipment schedules. Plumbing plans may also show a schedule of plumbing systems and plumbing system specifications.

2.9.0 Mechanical Plans

Mechanical plans show the heating, ventilating, and air conditioning systems, as well as other mechanical systems for a building. For some residential jobs, the mechanical plan may be combined with the plumbing plan and show very little detailed information other than the locations of the main HVAC system components. This is because the installation location of duct or piping runs is allowed to be determined on the job by the HVAC contractor. For large commercial jobs, the mechanical plans typically show information about the HVAC system installation and the installation of other equipment. Information about the installation of the plumbing and electrical systems is usually shown on separate stand-alone plumbing and electrical plan drawings.

As with separate plumbing plans, details of the mechanical plan are usually overlaid on tracings of the various building floor plans from which unnecessary details have been omitted to allow the location and layout of the HVAC system equipment to show clearly. *Figure 17* shows an example of a typical HVAC mechanical plan.

Mechanical plans typically contain tabulated schedules that identify the different items and types of HVAC equipment. As appropriate, detailed views describing the installation of the HVAC equipment are shown. Depending on the nature of the project, these views can include refrigeration piping schematics, chilled-water coil and/or hot-water coil piping schematics, and views detailing piping runs and pipe sizes for major items of HVAC equipment.

Mechanical plans also normally include an HVAC legend listing the various symbols pertaining to the plan. Common HVAC symbols are covered later in this module.

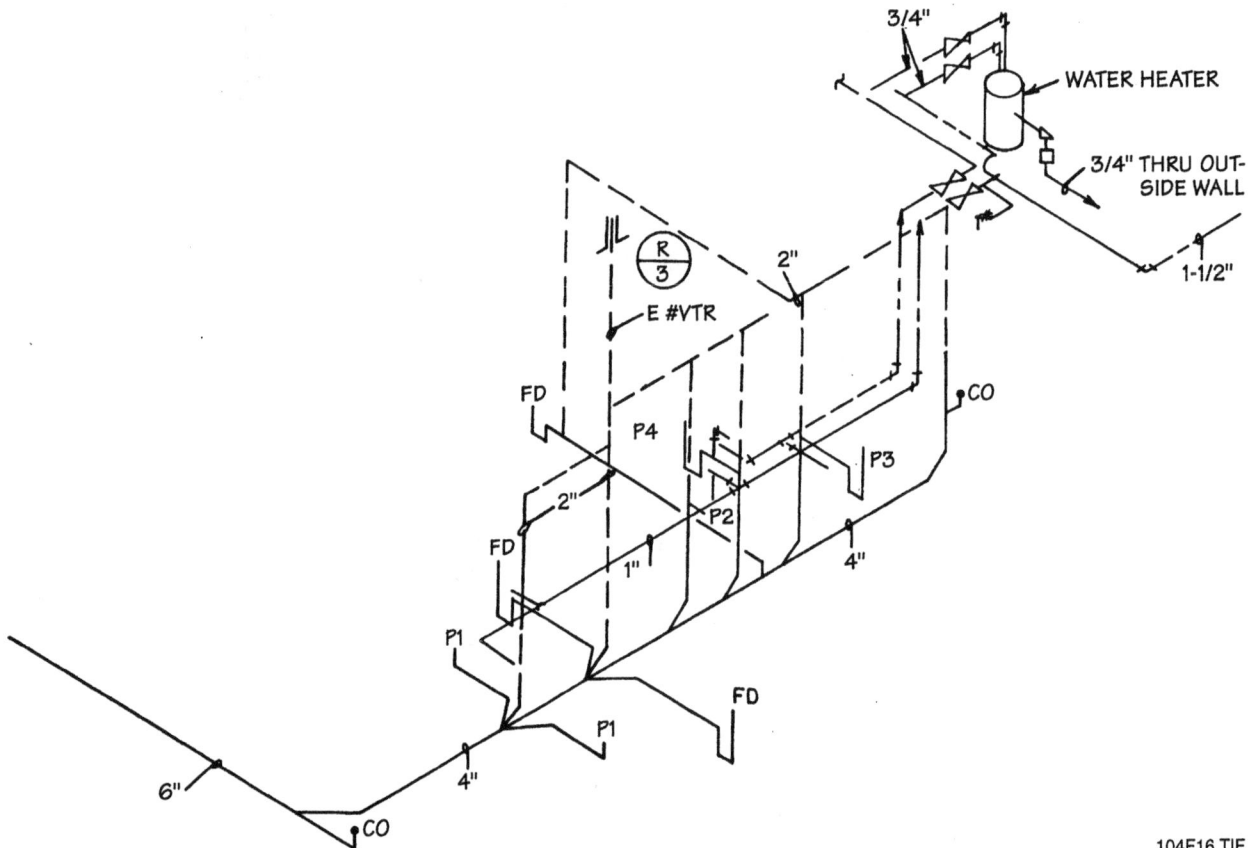

Figure 16 ◆ Plumbing riser diagram.

104F16.TIF

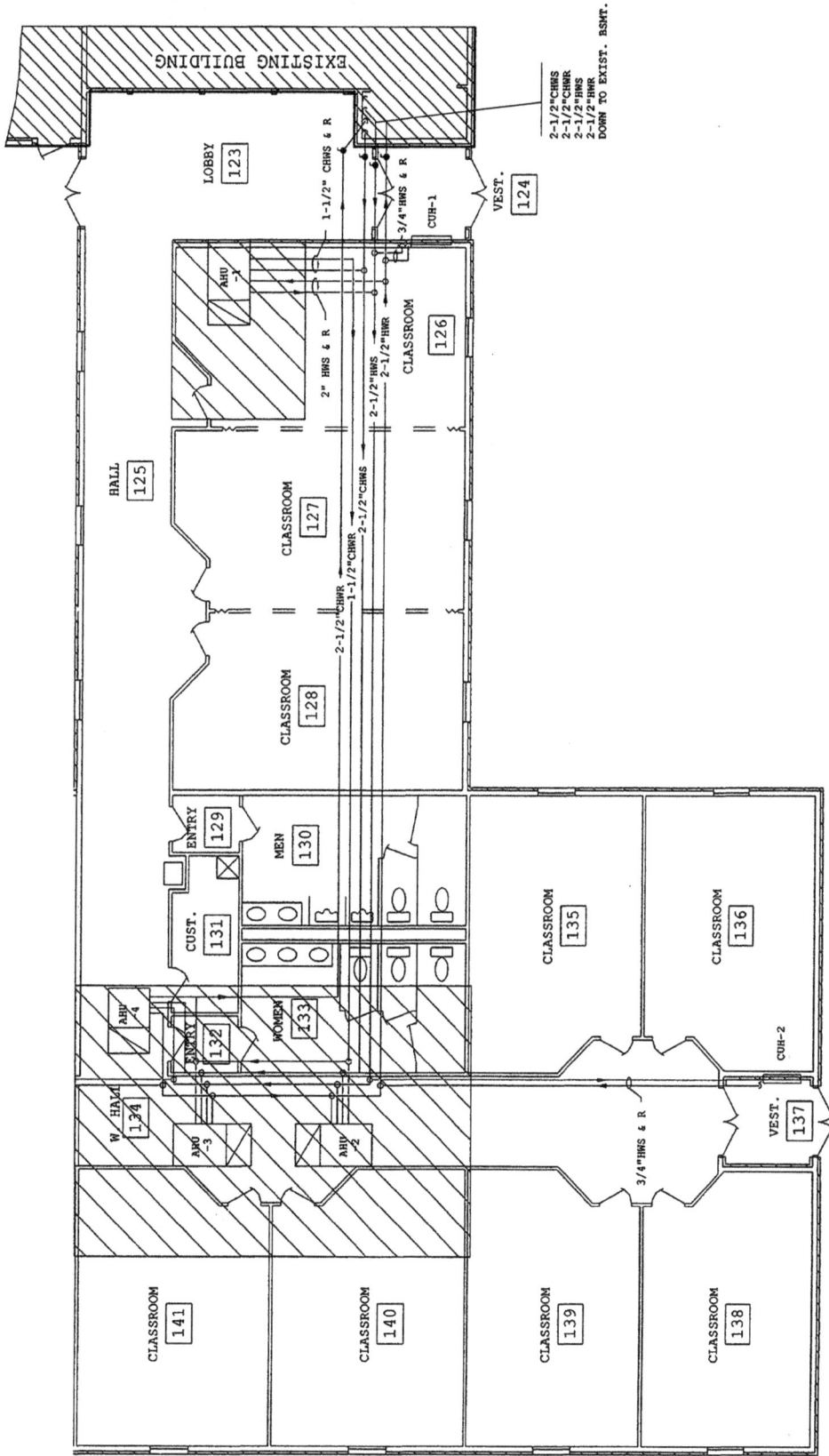

104F17A.TIF

Figure 17 ◆ Mechanical plan. (1 of 3)

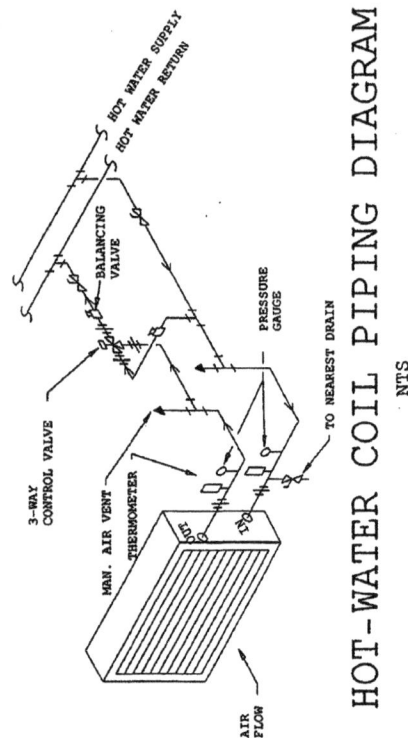

HOT-WATER COIL PIPING DIAGRAM
NTS

HOT WATER SUPPLY
HOT WATER RETURN
BALANCING VALVE
3-WAY CONTROL VALVE
MAN. AIR VENT
THERMOMETER
PRESSURE GAUGE
TO NEAREST DRAIN
IN
OUT
AIR FLOW

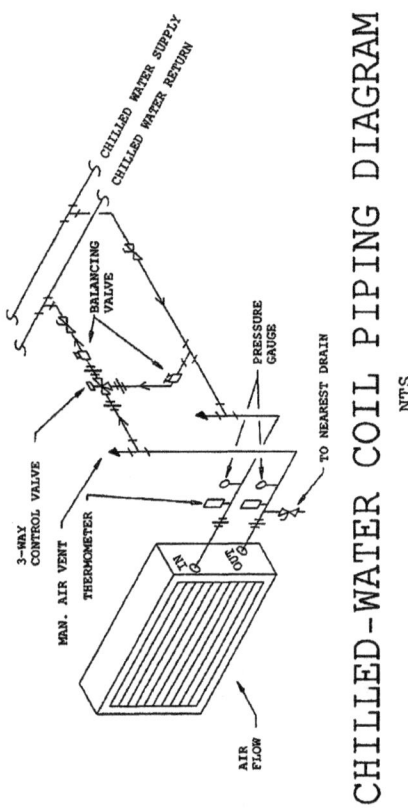

CHILLED-WATER COIL PIPING DIAGRAM
NTS

CHILLED WATER SUPPLY
CHILLED WATER RETURN
BALANCING VALVE
3-WAY CONTROL VALVE
MAN. AIR VENT
THERMOMETER
PRESSURE GAUGE
TO NEAREST DRAIN
IN
OUT
AIR FLOW

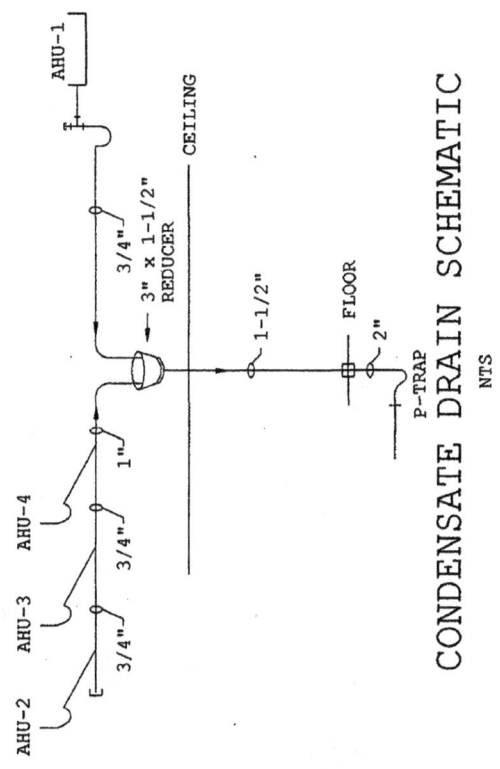

CONDENSATE DRAIN SCHEMATIC
NTS

AHU-1
3/4"
3" x 1-1/2" REDUCER
CEILING
1-1/2"
FLOOR
2"
P-TRAP
1"
3/4"
3/4"
AHU-4
AHU-3
AHU-2

104F17C.TIF

Figure 17 ◆ Mechanical plan. (2 of 3)

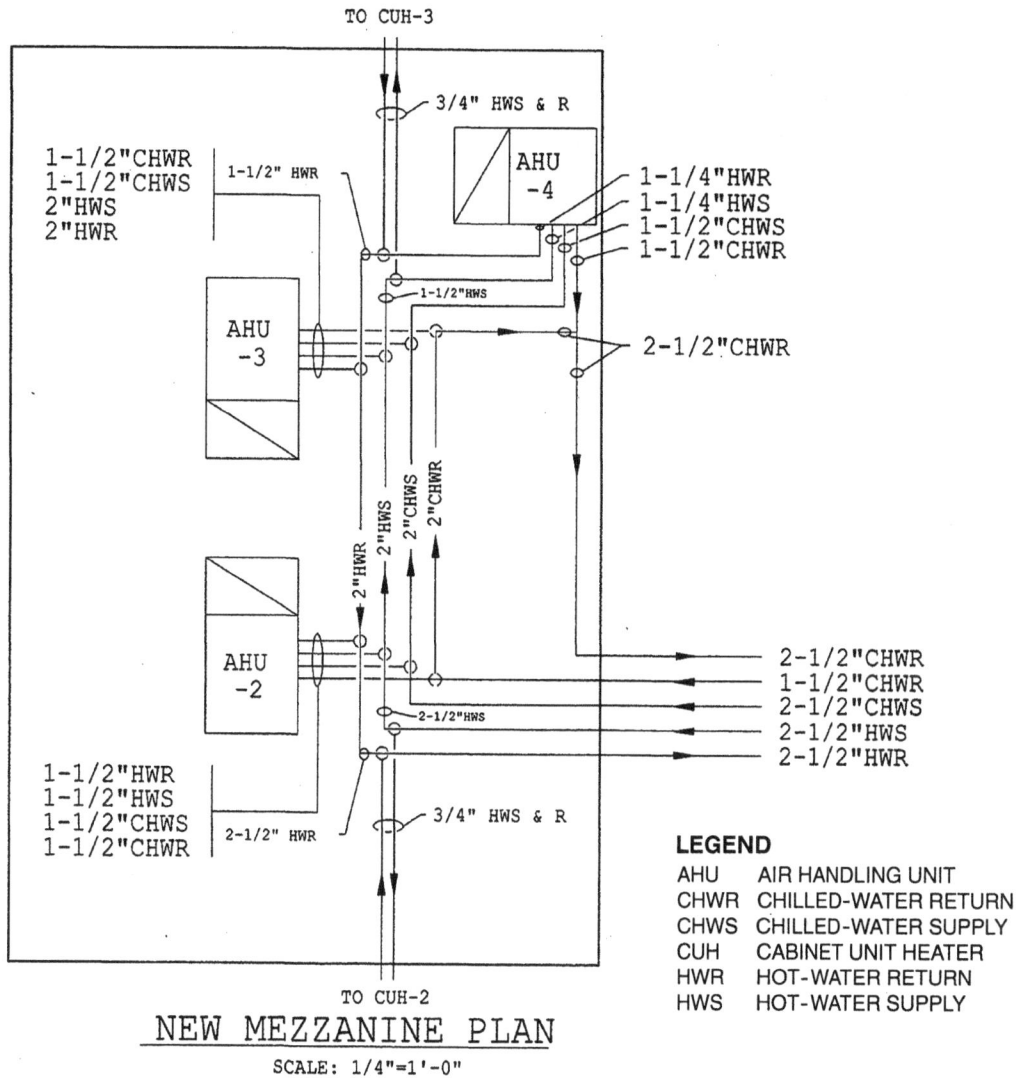

TO CUH-3

3/4" HWS & R

1-1/2"CHWR
1-1/2"CHWS
2"HWS
2"HWR

1-1/2" HWR

AHU
-4

1-1/4"HWR
1-1/4"HWS
1-1/2"CHWS
1-1/2"CHWR

1-1/2"HWS

AHU
-3

2-1/2"CHWR

2"HWR   2"HWS   2"CHWS   2"CHWR

AHU
-2

2-1/2"CHWR
1-1/2"CHWR
2-1/2"CHWS
2-1/2"HWS
2-1/2"HWR

2-1/2"HWS

1-1/2"HWR
1-1/2"HWS
1-1/2"CHWS
1-1/2"CHWR

2-1/2" HWR

3/4" HWS & R

TO CUH-2

NEW MEZZANINE PLAN

SCALE: 1/4"=1'-0"

LEGEND

AHU   AIR HANDLING UNIT
CHWR   CHILLED-WATER RETURN
CHWS   CHILLED-WATER SUPPLY
CUH   CABINET UNIT HEATER
HWR   HOT-WATER RETURN
HWS   HOT-WATER SUPPLY

104F17B.TIF

Figure 17 ◆ Mechanical plan. (3 of 3)

2.10.0 Electrical Plans

For smaller construction jobs, the electrical plans are usually shown on the architectural floor plans. For large commercial jobs, the electrical plans are typically stand-alone drawings that show only information about the electrical system installation. Like the separate plumbing and mechanical plans, electrical plans typically overlay tracings of the various building floor plans from which unnecessary details have been omitted to allow the location and layout of the electrical system equipment to show clearly. *Figure 18* shows an example of a typical electrical plan.

Electrical plans show the locations of the meter, distribution panels, light fixtures, switches, and other electrical equipment. Also shown are equipment and fixtures schedules, an electrical legend listing the various symbols pertaining to the plan, specifications for load capacities, and wire sizes. Electrical plans usually have a power riser diagram that shows all the major pieces of electrical equipment, including HVAC equipment, as well as the connecting lines used to indicate service-entrance conductors and feeders. Electrical plans may also contain information about the electrical specifications.

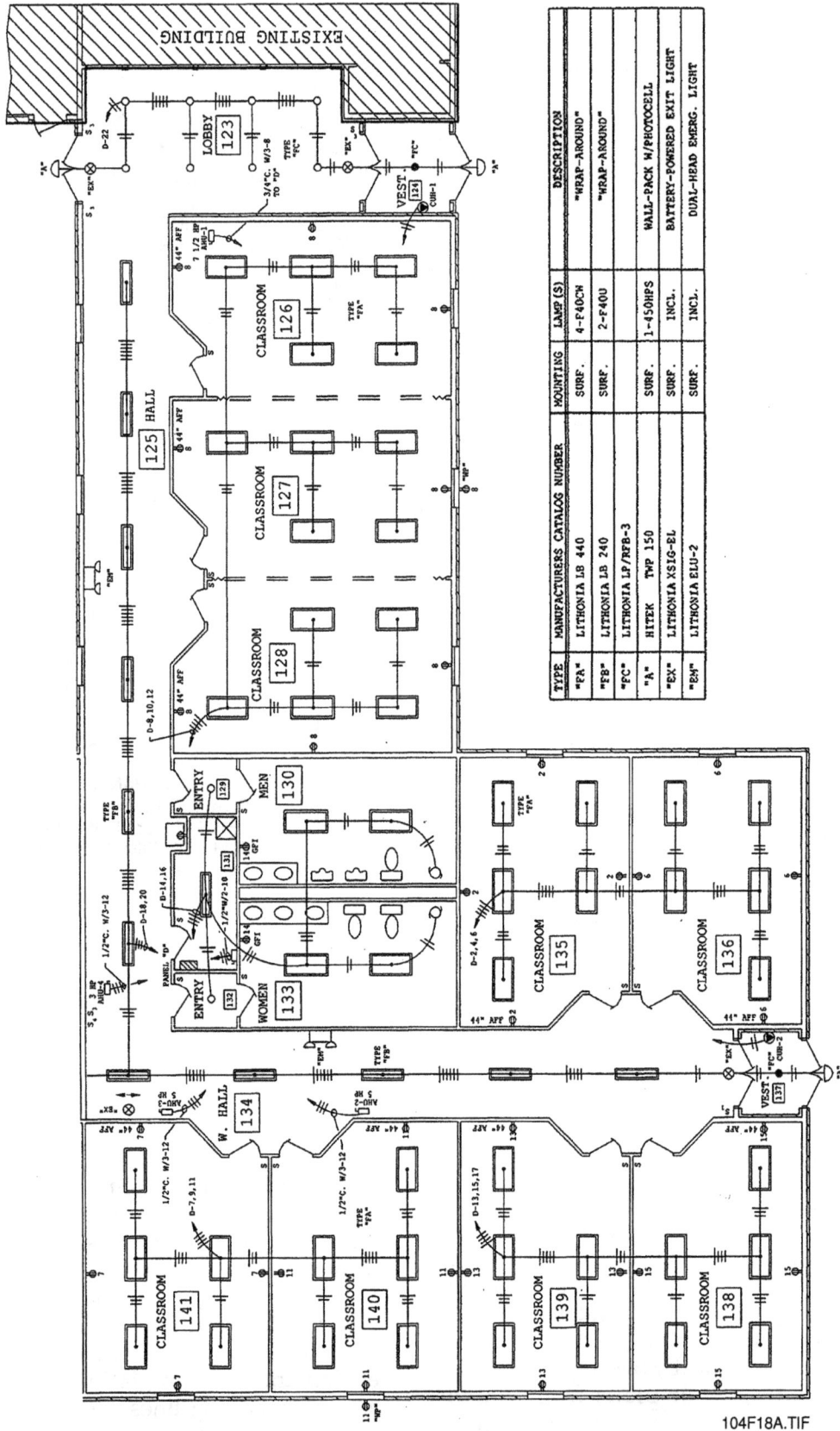

The electrical plan table contains:

TYPE	MANUFACTURERS CATALOG NUMBER	MOUNTING	LAMP(S)	DESCRIPTION
"A"	LITHONIA LB 440	SURF.	4-F40CW	"WRAP-AROUND"
"B"	LITHONIA LB 240	SURF.	2-F40U	"WRAP-AROUND"
"C"	LITHONIA LP/RFB-3			
"A"	HITEK TWP 150	SURF.	1-450HPS	WALL-PACK W/PHOTOCELL
"EX"	LITHONIA XSIG-EL	SURF.	INCL.	BATTERY-POWERED EXIT LIGHT
"EM"	LITHONIA ELU-2	SURF.	INCL.	DUAL-HEAD EMERG. LIGHT

104F18A.TIF

Figure 18 ◆ Electrical plan. (1 of 2)

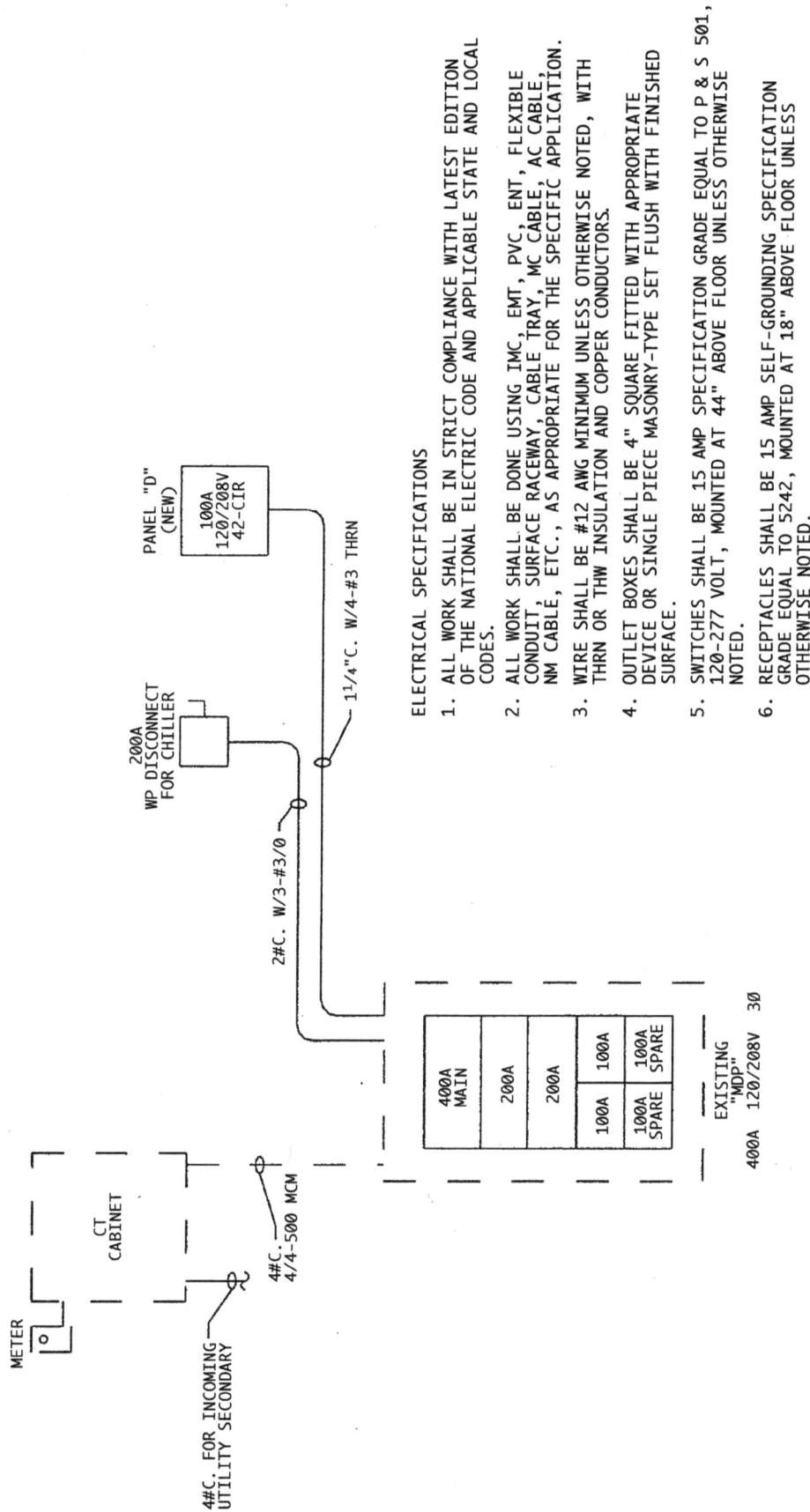

PANEL "D"
(NEW)

100A
120/208V
42-CIR

1¼"C. W/4-#3 THRN

200A
WP DISCONNECT
FOR CHILLER

2#C. W/3-#3/0

METER

CT
CABINET

4#C.
4/4-500 MCM

4#C. FOR INCOMING
UTILITY SECONDARY

400A
MAIN

200A

200A

100A | 100A

100A
SPARE | 100A
SPARE

EXISTING
"MDP"

400A   120/208V   3Ø

RISER DIAGRAM

NTS

ELECTRICAL SPECIFICATIONS

1. ALL WORK SHALL BE IN STRICT COMPLIANCE WITH LATEST EDITION OF THE NATIONAL ELECTRIC CODE AND APPLICABLE STATE AND LOCAL CODES.

2. ALL WORK SHALL BE DONE USING IMC, EMT, PVC, ENT, FLEXIBLE CONDUIT, SURFACE RACEWAY, CABLE TRAY, MC CABLE, AC CABLE, NM CABLE, ETC., AS APPROPRIATE FOR THE SPECIFIC APPLICATION.

3. WIRE SHALL BE #12 AWG MINIMUM UNLESS OTHERWISE NOTED, WITH THRN OR THW INSULATION AND COPPER CONDUCTORS.

4. OUTLET BOXES SHALL BE 4" SQUARE FITTED WITH APPROPRIATE DEVICE OR SINGLE PIECE MASONRY-TYPE SET FLUSH WITH FINISHED SURFACE.

5. SWITCHES SHALL BE 15 AMP SPECIFICATION GRADE EQUAL TO P & S 501, 120-277 VOLT, MOUNTED AT 44" ABOVE FLOOR UNLESS OTHERWISE NOTED.

6. RECEPTACLES SHALL BE 15 AMP SELF-GROUNDING SPECIFICATION GRADE EQUAL TO 5242, MOUNTED AT 18" ABOVE FLOOR UNLESS OTHERWISE NOTED.

7. RECEPTACLES, SWITCH, TELEPHONE, ETC., COVER PLATES SHALL BE SMOOTH IVORY PLASTIC EQUAL TO SIERRA "P" SERIES.

8. MATERIAL AND EQUIPMENT SHALL BE NEW, OF STANDARD MANUFACTURER'S CONSTRUCTION, INSTALLED IN ACCORDANCE WITH ACCEPTED PRACTICE BY COMPETENT WORKERS.

104F18B.TIF

Figure 18 ◆ Electrical plan. (2 of 2)

2.11.0 Shop Drawings

Shop drawings are specialized drawings that show how to fabricate and install components of a construction project. One type of shop drawing that a drafter creates after the engineer designs the structure is a detail drawing. It shows the locations of all holes and openings and provides notes specifying how the part is to be made. Assembly instructions may be included. This type of drawing is used principally for structural steel members. Shop drawings related to structural steel members are covered in more detail later in this module.

Another type of shop drawing (or submittal) pertains to the purchase of special items of equipment for installation in a building. This kind of shop drawing is usually prepared by the equipment manufacturer. This drawing shows overall sizes, details of construction, methods of securing the equipment to the structure, and all pertinent data that the architect and contractor need to know for the final placement and installation of the equipment.

Shop drawings produced by a contractor or **fabricator** are usually submitted to the owner or architect for approval and revisions or corrections. The design drawing is often put on the same sheet as the shop drawing.

2.12.0 As-Built Drawings

As-built drawings are drawings that are formally incorporated into the drawing set to record changes. These drawings are marked up by the various trades to show any differences between what was originally shown on a plan by the architect or engineer and what was actually built. Such changes result from the need to relocate equipment to avoid obstructions; to alter the location of a door, window, wall, etc., for some reason; or because the architect has changed a certain detail in the building design in response to customer preferences. On many jobs, any such changes to the design can only be made after an RFI or **change order** has been generated and approved by the architect or other designated person. Depending on the complexity of the change, as-built drawings are typically outlined with a unique design or marked in red ink to make sure they stand out. Changes must be dated and initialed by the responsible party.

2.13.0 Soil Reports

Soil conditions are one of the factors that determine the type of foundation best suited for a structure. Building a structure on soil where the soil conditions can cause a large amount of uneven settlement to occur can result in cracks in the foundation and structural damage to the rest of the building. Therefore, in designing the foundation for a structure, an architect must consider the soil conditions of the building site. Typically, the architect consults a soil engineer who makes test bores of the soil on the building site and analyzes the samples. The results of the soil analysis are summarized in a soil report issued by the engineer. This report is often included as part of the drawing set. Consider all aspects of the soil report, including elevation of the water table.

3.0.0 ◆ READING AND INTERPRETING DRAWINGS

In order to read and interpret the information on drawings, you need to learn the special language used in construction drawings. This section of the module describes the different types of lines, dimensioning, symbols, and abbreviations used on drawings.

3.1.0 Site/Plot Plans

Site/plot plans show the positions and sizes of all relevant structures on the site as well as the features of the terrain. It is difficult to show the amount of information required on these drawings if symbols are not used, especially drawings that are nonstandard. *Figure 19* shows some symbols commonly used on site/plot diagrams.

Note that the contour line symbols shown on *Figure 19* are used to show changes in the elevation and contour of the land. The lines may be dashed or solid. Generally, dashed lines are used to show the natural or existing grade, and solid lines show the finished grade to be achieved during construction.

Each contour line across the plot of land represents a line of constant elevation relative to some point such as sea level or a local feature. *Figure 20* shows an example of a contour map for a hill. As shown, contour lines are drawn in uniform elevation intervals called contour intervals. Commonly used intervals are 1', 2', 5', and 10'.

On some plans and surveys, every fifth contour line is drawn using a heavier-weight line and is labeled with its elevation to help the user more easily determine the contour. This method of drawing contour lines is called indexing contours. The elevation is marked above the contour line, or the line is interrupted for it.

As shown in *Figure 20*, contour lines form a closed loop within the map. If you start at any point on the contour and follow its path, you will eventually return to the starting point. A contour

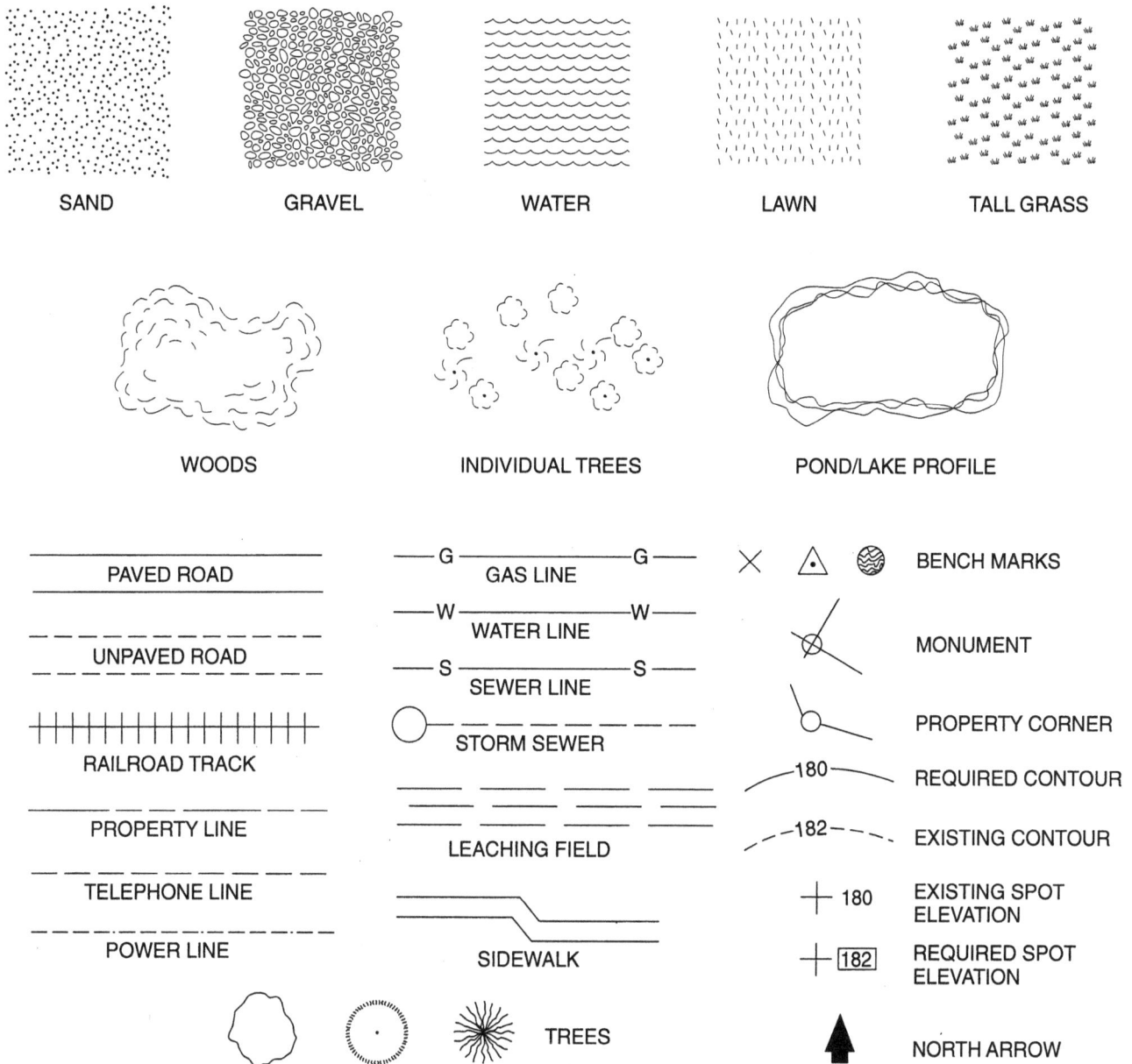

SAND   GRAVEL   WATER   LAWN   TALL GRASS

WOODS   INDIVIDUAL TREES   POND/LAKE PROFILE

PAVED ROAD

UNPAVED ROAD

RAILROAD TRACK

PROPERTY LINE

TELEPHONE LINE

POWER LINE

GAS LINE

WATER LINE

SEWER LINE

STORM SEWER

LEACHING FIELD

SIDEWALK

TREES

BENCH MARKS

MONUMENT

PROPERTY CORNER

REQUIRED CONTOUR

EXISTING CONTOUR

EXISTING SPOT ELEVATION

REQUIRED SPOT ELEVATION

NORTH ARROW

104F19.EPS

Figure 19 ◆ Common site/plot plan symbols.

may close on a site plan or map, or it may be discontinued at any two points at the borders of the plan or map. Examples of this are shown on the site plan shown earlier in this module. Such points mark the ends of the contour on the map, but the contour does not end at these points. The contour is continued on a plan or map of the adjacent land. Some rules for interpreting contours include the following:

- Contour lines do not cross.
- Contour lines crossing a stream point upstream.

- The horizontal distance between contour lines represents the degree of slope. Closely spaced contour lines represent steep ground, and widely spaced contour lines represent nearly level ground with a gradual slope. Uniform spacing indicates a uniform slope.
- Contour lines are at right angles to the slope. Therefore, water flow is perpendicular to contour lines.
- Straight contour lines parallel to each other represent man-made features such as terracing.

308

300

275

STREAM

300

275

104F20.EPS

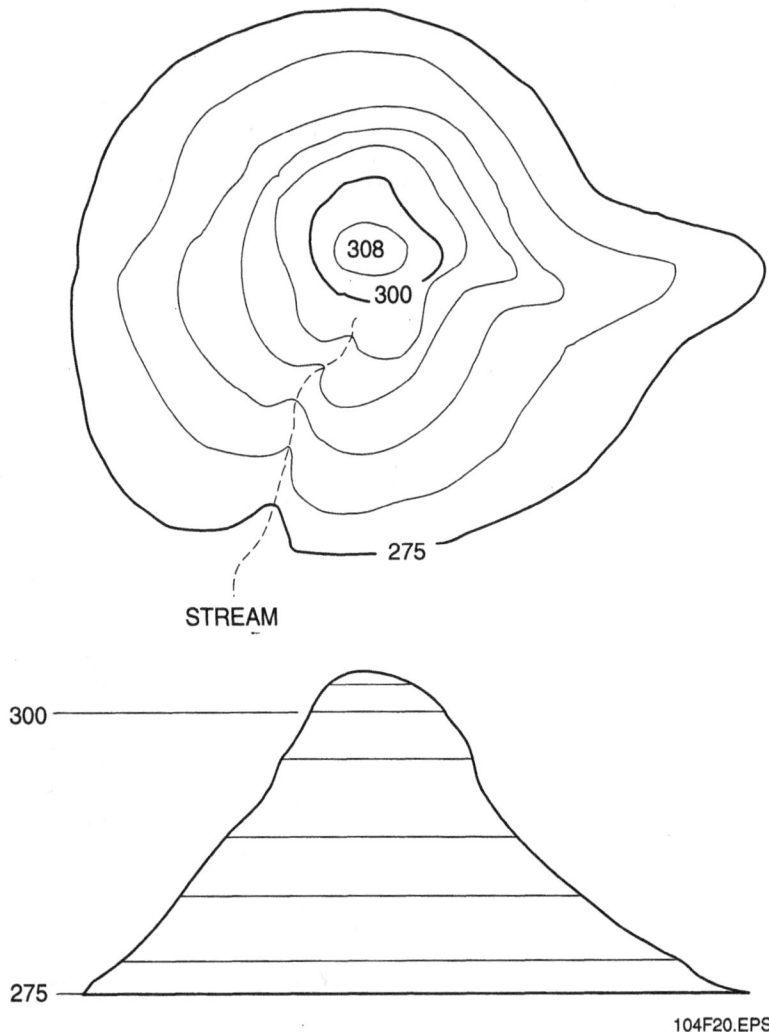

Figure 20 ◆ Contour map of a hill.

3.2.0 Lines Used on Drawings

Many different types of lines are used to draw and describe a structure. Lines are drawn wide, narrow, dark, light, broken, and unbroken, with each type of line conveying a specific meaning. *Figure 21* shows the most common lines used on construction drawings. The description of each type of line is as follows:

- *Object lines* – Heavier-weight lines used to show the main outline of the structure, including exterior walls, interior partitions, porches, patios, sidewalks, parking lots, and driveways.
- *Dimension and extension lines* – Provide the dimensions of an object. An extension line is drawn out from an object at both ends of the part to be measured. Extension lines are not supposed to touch the object lines. This is so

they cannot be confused with the object lines. A dimension line is drawn at right angles between the extension lines and a number placed above, below, or to the side of it to indicate the length of the dimension line. Sometimes a gap is made in the dimension line and the number is written in the gap.

- *Leader line* – Connects a note or dimension to a related part of the drawing. Leader lines are usually curved or at an angle from the feature being distinguished to avoid confusion with dimension and other lines.
- *Center line* – Designates the center of an area or object and provides a reference point for dimensioning. Center lines are typically used to indicate the centers of objects such as columns, posts, footings, and door openings.

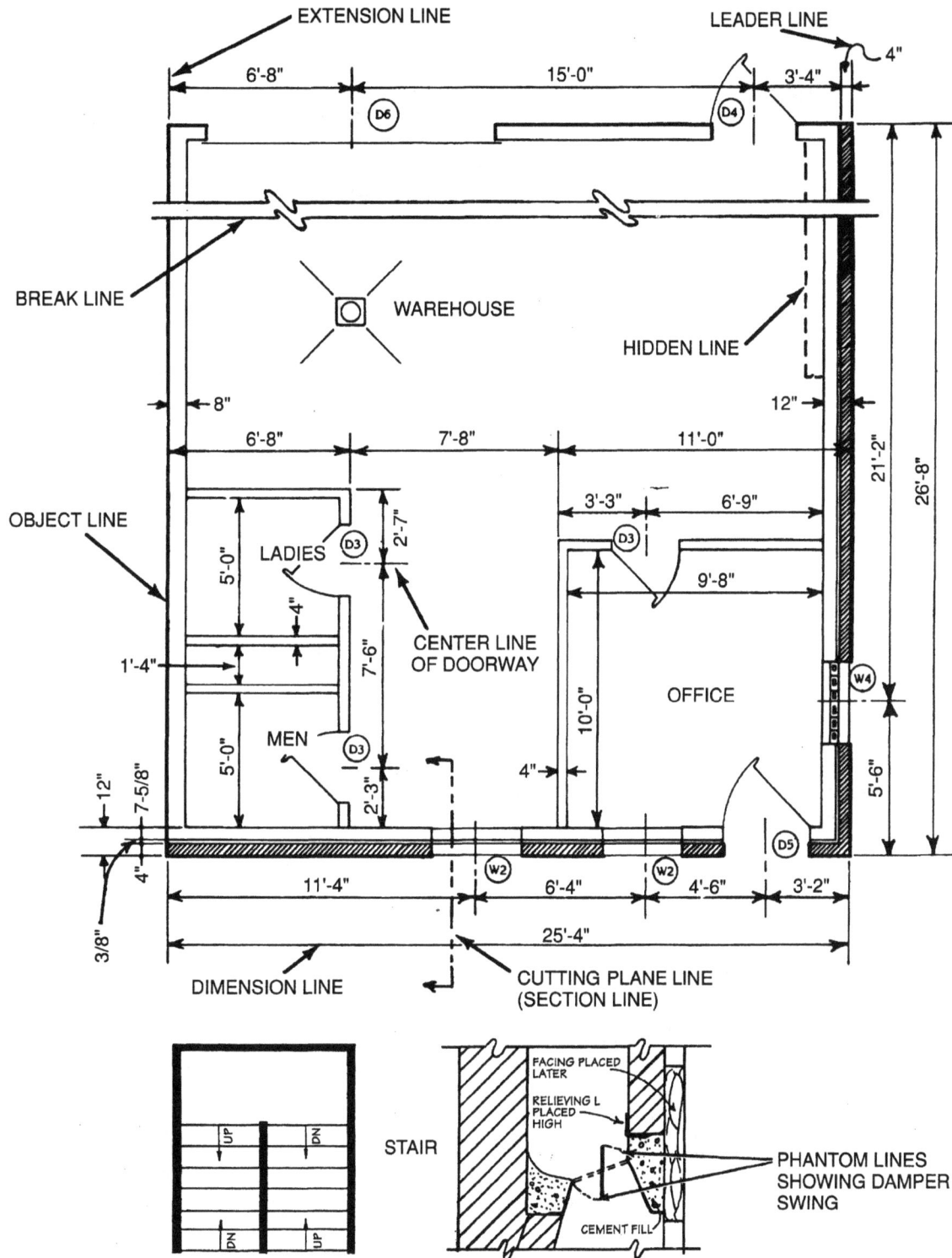

Figure 21 ◆ Drawing lines.

- *Cutting plane (section line)* – Indicates an area that has been cut away and shown in a section view so that the interior features can be seen. The arrows at the ends of the cutting plane indicate the direction in which the section is viewed. Letters identify the cross-sectional view of that specific part of the structure. More elaborate methods of labeling section reference lines are used in larger, more complicated sets of plans (*Figure 22*). The sectional drawing may be on the same page as the reference line or on another page.

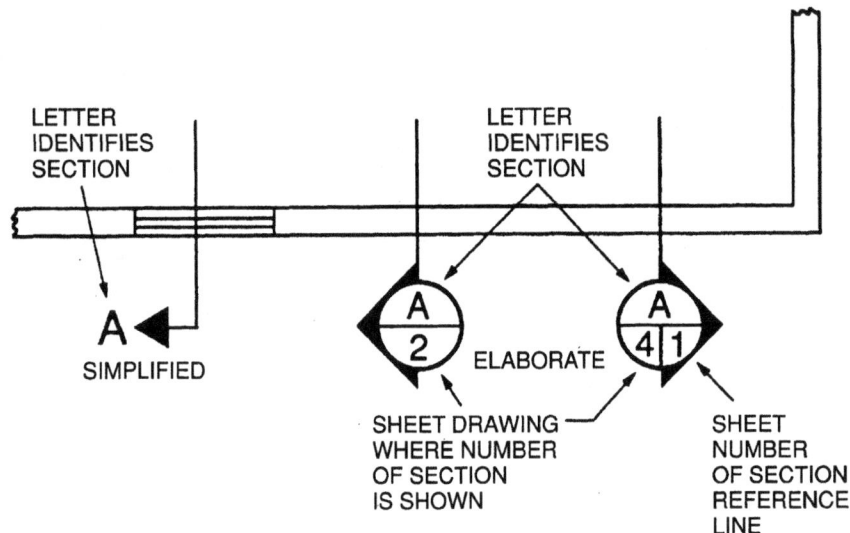

LETTER
IDENTIFIES
SECTION

LETTER
IDENTIFIES
SECTION

A
SIMPLIFIED

A
2
ELABORATE

A 4|1

SHEET DRAWING
WHERE NUMBER
OF SECTION
IS SHOWN

SHEET
NUMBER
OF SECTION
REFERENCE
LINE

104F22.TIF

Figure 22 ◆ Methods of labeling section reference lines.

- *Break line* – Shows that an object or area has not been drawn in its entirety.
- *Hidden line* – Indicates an outline that is invisible to an observer because it is covered by another surface or object that is closer to the observer.
- *Phantom line* – Indicates alternative positions of moving parts, such as a damper's swing, or adjacent positions of related parts. It may also be used to represent repeated details.
- *Stair indicator line* – A short line with an arrowhead that shows the ascent or descent of stairs on a floor plan.

3.3.0 Symbols Used on Drawings

Symbols are used in architectural plans and drawings to pictorially show different kinds of materials, fixtures, and structural members. The meanings of symbols and the types used are not standardized and can vary from location to location. A set of drawings generally includes a sheet that identifies the specific symbols used and their meanings. When using any drawing set, you should always refer to this sheet of symbols in order to avoid making mistakes when reading the drawings.

Some symbols are not intended to show part of an object, but are provided to guide you through the drawings themselves. Examples of this type include door and window designators that refer to door and window schedules where the different types are described. Other symbols are used to show the orientation of the object, showing the direction or side (north, south, front, back, and so on).

For electrical, plumbing, and HVAC, there are symbols that show the types of equipment to be installed, such as switches, lavatories, and warm air supply ducts. Each trade has its own symbols and abbreviations.

Symbols used on drawings generally fall into the following categories:

- Material (general plan) symbols
- Window and door symbols
- Electrical symbols
- Plumbing symbols
- HVAC symbols
- Structural member symbols
- Welding symbols

3.3.1 Material Symbols

Typically, materials are shown on drawings in two ways (*Figure 23*). One way they are shown is as a plan view, where the material is presented as you would see it when looking down on it. The second way materials are shown is as an elevation view or section view, where the symbol shows the material roughly as it looks when you are facing it.

3.3.2 Window and Door Symbols

Floor plans show a great deal of information in a small drawing area. Therefore, you have to rely on symbols for much of the information. Window and door symbols are usually shown with only a center line measurement to locate them from some reference point. However, more information about windows and doors can be found in separate window and door schedules contained in the drawing set. Window and door symbols are shown on drawings as either elevation or plan views (*Figure 24*).

WOOD
Finish   Rough   Wood Finish on Stud

CONCRETE
Concrete   Block

INSULATION
Loose Fill or Batts   Boards or Quilts

GLASS
Block   Sheet & Plate

BRICK
Common   Face   Fire Brick on Common   Brick Veneer

METAL
Steel, Iron   Structural Steel   Brass, Bronze   Metal Stud Partition with Drywall
Aluminum

STONE
Cut Stone   Rubble   Marble   Slate, Bluestone, Soapstone

MISCELLANEOUS
Waterproofing, Felt, Flashing, Etc.   Plaster, Sand, Cement & Drywall   Reinforcing Bars   Glazed Clay Tile
Earth   Gravel   Gravel with Sand   Cinders

GENERAL PLAN SYMBOLS

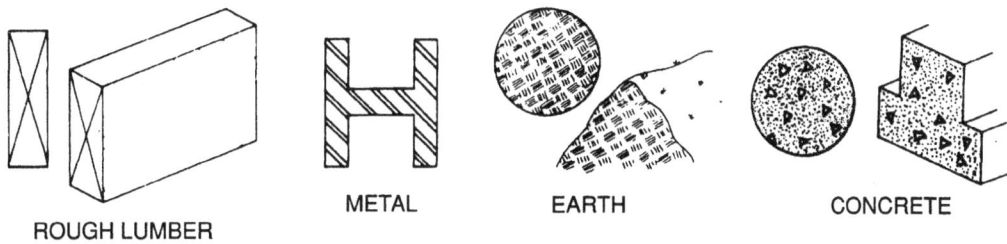

ROUGH LUMBER   METAL   EARTH   CONCRETE

SECTION VIEW SYMBOLS

104F23.TIF

Figure 23 ◆ Material symbols.

DOOR TYPE	SYMBOL	WINDOW TYPE	SYMBOL
SINGLE SWING		AWNING	
SLIDER		FIXED SASH	
BIFOLD		DOUBLE HUNG	
FRENCH		CASEMENT	
ACCORDION		HORIZONTAL SLIDER	

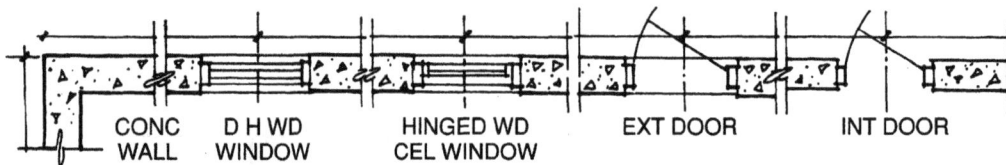

CONC WALL D H WD WINDOW HINGED WD CEL WINDOW EXT DOOR INT DOOR

DOORS AND WINDOWS IN CONCRETE WALL

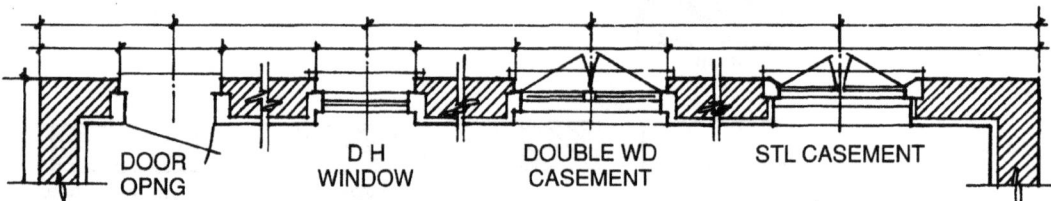

DOOR OPNG D H WINDOW DOUBLE WD CASEMENT STL CASEMENT

DOORS AND WINDOWS IN SOLID BRICK WALL

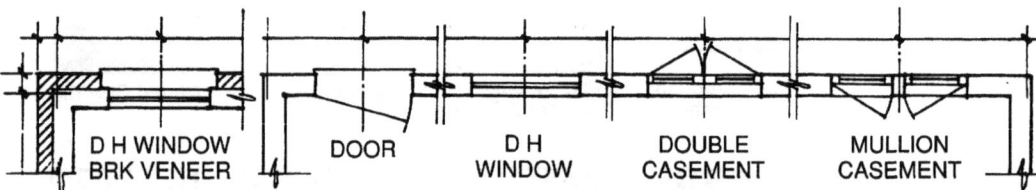

D H WINDOW BRK VENEER DOOR D H WINDOW DOUBLE CASEMENT MULLION CASEMENT

DOORS AND WINDOWS IN BRICK VENEER AND WOOD STUD WALLS

104F24.TIF

Figure 24 ◆ Window and door symbols.

3.3.3 Electrical Symbols

Electrical symbols are used to show the electrician the locations of outlets and switches. They are also used to show auxiliary hardware such as buzzers, telephones, and computer hookups. *Figure 25* shows some common electrical symbols.

Switching arrangements are also indicated on the floor plans by using symbols. *Figure 26* shows some symbolic notations commonly used with electrical symbols on a floor plan.

Figure 26A shows the electrician that two outlets are controlled by two three-way switches.

Similarly, *Figure 26B* shows that two outlets are controlled by two three-way switches and a four-way switch.

Figures 26C through *26F* show various outlets and fixtures controlled by a single-pole switch. (A single-pole switch is a device that opens and closes one side of the circuit only.)

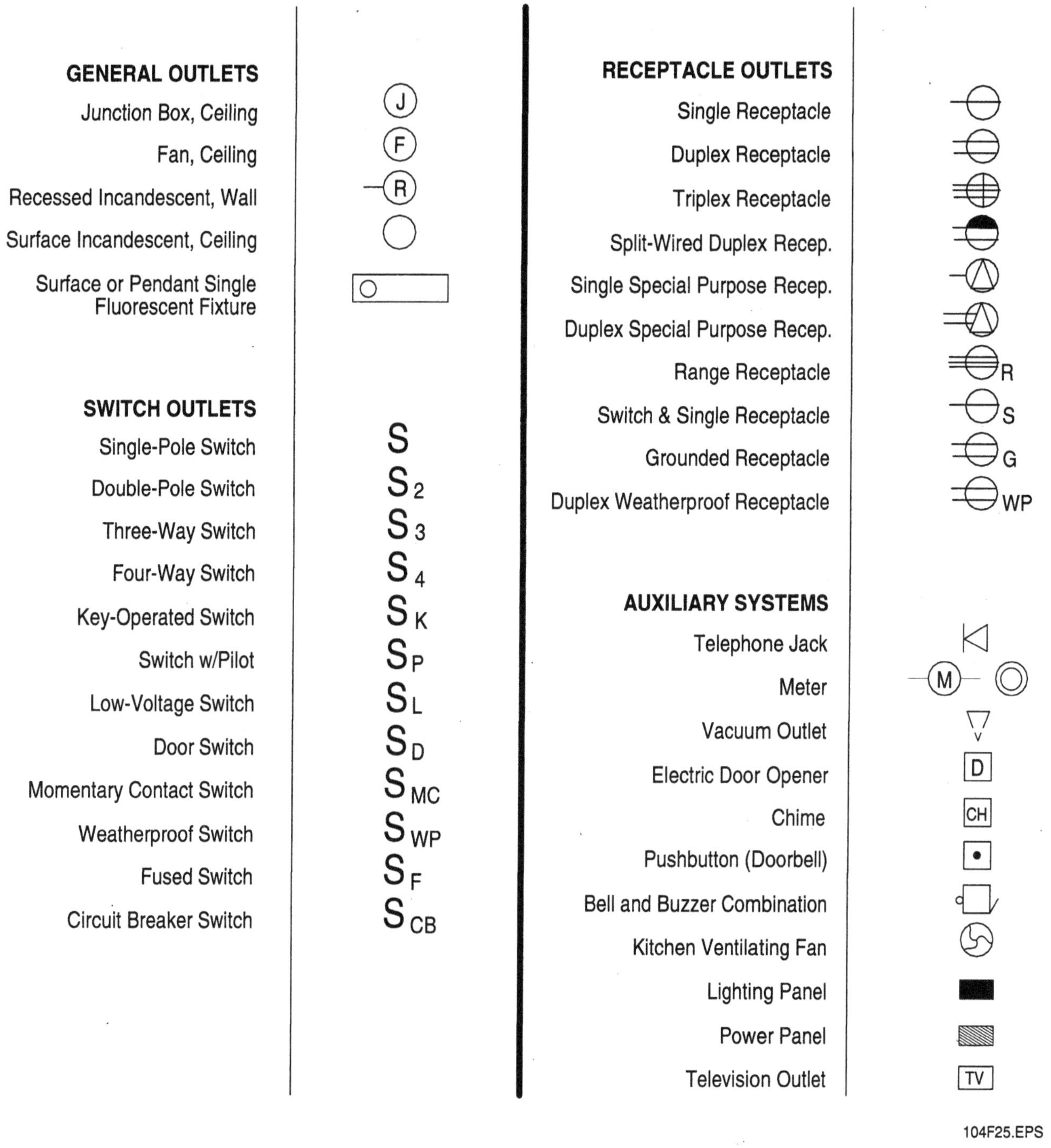

GENERAL OUTLETS

Junction Box, Ceiling   ⓙ

Fan, Ceiling   Ⓕ

Recessed Incandescent, Wall   —Ⓡ

Surface Incandescent, Ceiling   ◯

Surface or Pendant Single
Fluorescent Fixture   [◯]

SWITCH OUTLETS

Single-Pole Switch S

Double-Pole Switch S_2

Three-Way Switch S_3

Four-Way Switch S_4

Key-Operated Switch S_K

Switch w/Pilot S_P

Low-Voltage Switch S_L

Door Switch S_D

Momentary Contact Switch S_{MC}

Weatherproof Switch S_{WP}

Fused Switch S_F

Circuit Breaker Switch S_{CB}

RECEPTACLE OUTLETS

Single Receptacle

Duplex Receptacle

Triplex Receptacle

Split-Wired Duplex Recep.

Single Special Purpose Recep.

Duplex Special Purpose Recep.

Range Receptacle   R

Switch & Single Receptacle   S

Grounded Receptacle   G

Duplex Weatherproof Receptacle   WP

AUXILIARY SYSTEMS

Telephone Jack

Meter   —Ⓜ— ◯

Vacuum Outlet

Electric Door Opener   [D]

Chime   [CH]

Pushbutton (Doorbell)   [•]

Bell and Buzzer Combination

Kitchen Ventilating Fan

Lighting Panel   ■

Power Panel

Television Outlet   [TV]

104F25.EPS

Figure 25 ◆ Electrical symbols.

3.3.4 Plumbing Symbols

Plumbing symbols show all of the hardware and fixtures required for the building. Floor plans usually indicate only where the different fixtures should be located and plumbed.

Figure 27 shows some common fixture symbols as they would appear on a floor plan. It also shows some common piping symbols that would be used on special plumbing diagrams.

3.3.5 HVAC Symbols

The locations and types of HVAC equipment are shown on the floor plan(s), sections, and detail drawings using symbols. For HVAC systems, symbols are also used to show the direction of movement of the hot and cold air in the system. *Figure 28* shows some common HVAC symbols.

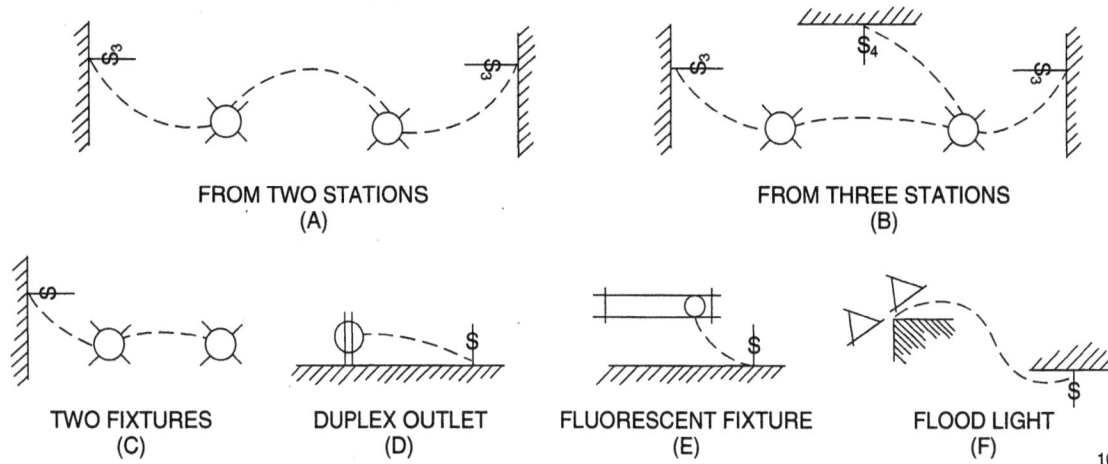

FROM TWO STATIONS
(A)

FROM THREE STATIONS
(B)

TWO FIXTURES
(C)

DUPLEX OUTLET
(D)

FLUORESCENT FIXTURE
(E)

FLOOD LIGHT
(F)

104F26.EPS

Figure 26 ◆ Electrical symbols showing control of an outlet.

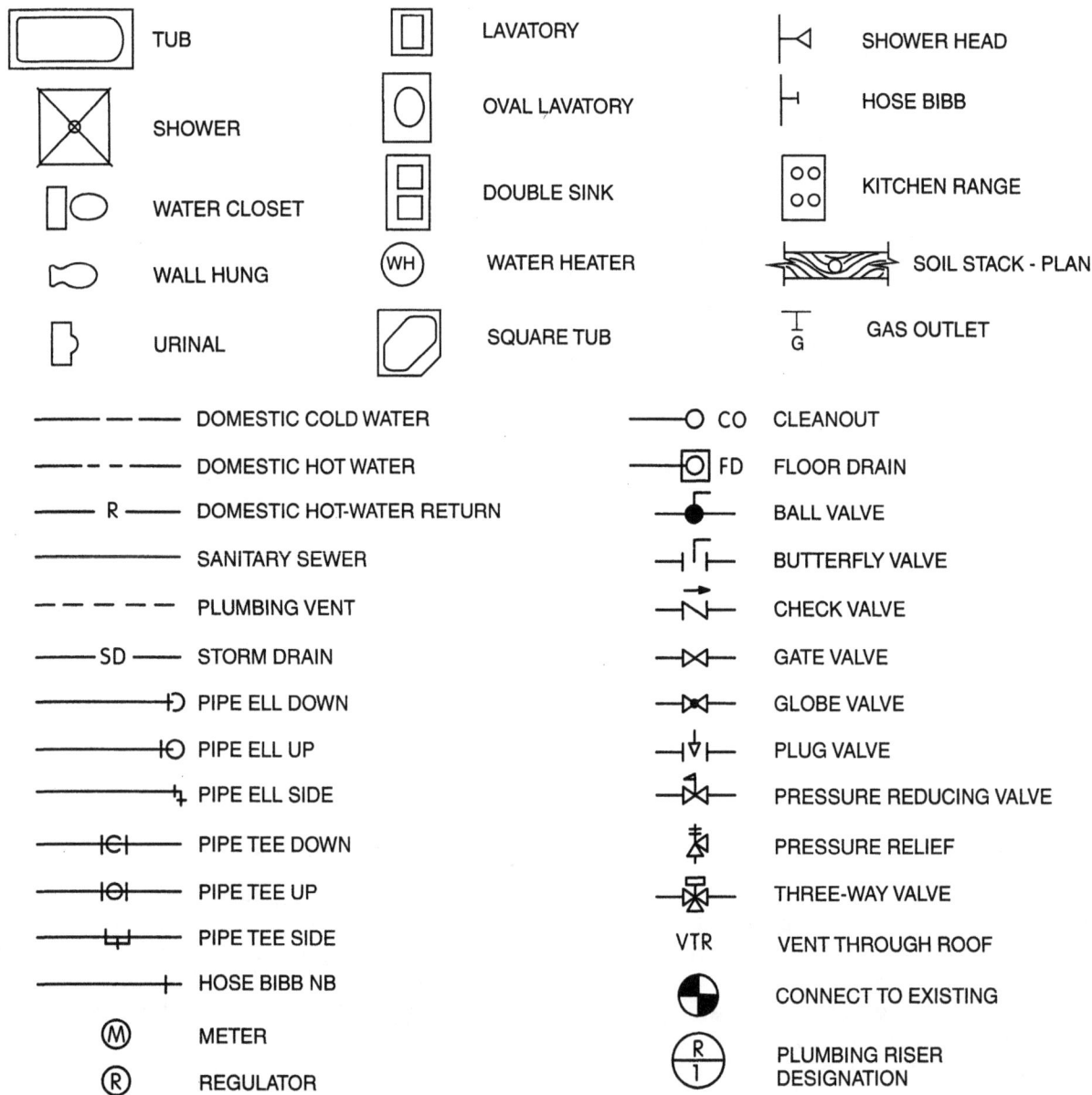

TUB

SHOWER

WATER CLOSET

WALL HUNG

URINAL

LAVATORY

OVAL LAVATORY

DOUBLE SINK

WATER HEATER

SQUARE TUB

SHOWER HEAD

HOSE BIBB

KITCHEN RANGE

SOIL STACK - PLAN

GAS OUTLET

DOMESTIC COLD WATER

DOMESTIC HOT WATER

DOMESTIC HOT-WATER RETURN

SANITARY SEWER

PLUMBING VENT

STORM DRAIN

PIPE ELL DOWN

PIPE ELL UP

PIPE ELL SIDE

PIPE TEE DOWN

PIPE TEE UP

PIPE TEE SIDE

HOSE BIBB NB

METER

REGULATOR

CLEANOUT

FLOOR DRAIN

BALL VALVE

BUTTERFLY VALVE

CHECK VALVE

GATE VALVE

GLOBE VALVE

PLUG VALVE

PRESSURE REDUCING VALVE

PRESSURE RELIEF

THREE-WAY VALVE

VENT THROUGH ROOF

CONNECT TO EXISTING

PLUMBING RISER
DESIGNATION

104F27.EPS

Figure 27 ◆ Plumbing symbols.

4.32

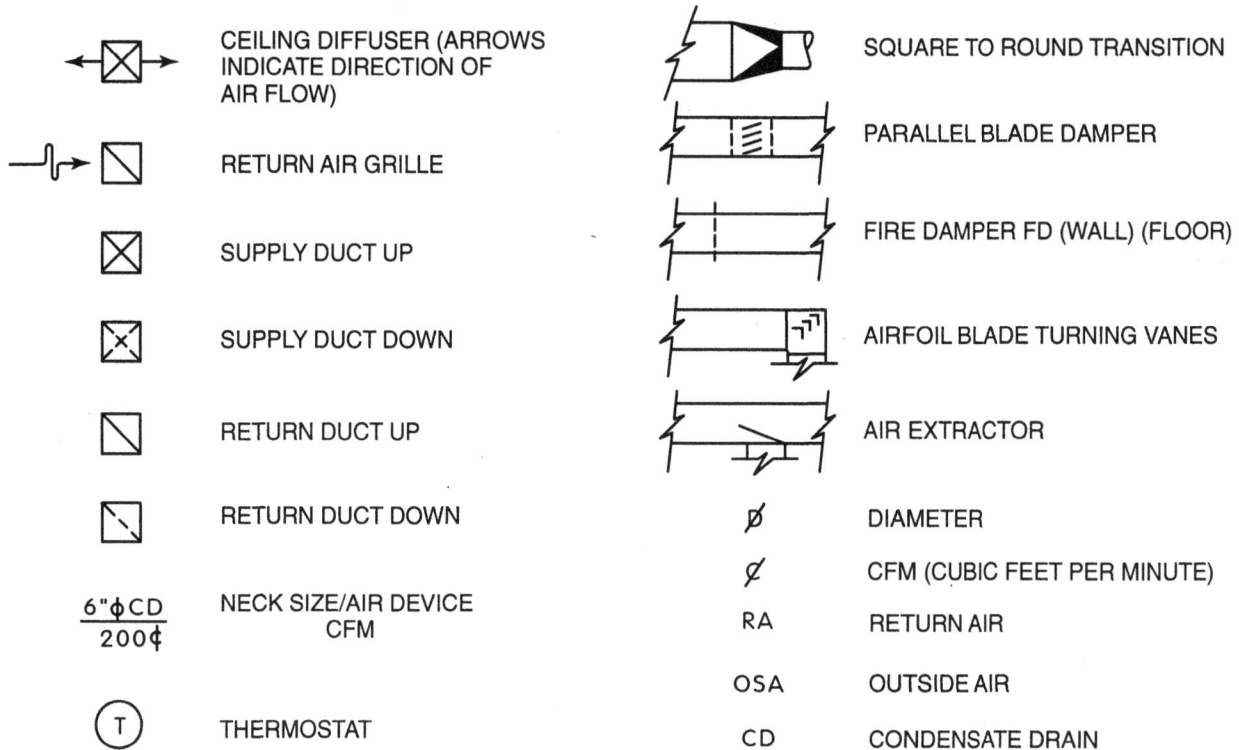

⊠ (arrows)	CEILING DIFFUSER (ARROWS INDICATE DIRECTION OF AIR FLOW)
	RETURN AIR GRILLE
⊠	SUPPLY DUCT UP
⊠	SUPPLY DUCT DOWN
◻	RETURN DUCT UP
◻	RETURN DUCT DOWN
6"ϕCD / 200¢	NECK SIZE/AIR DEVICE CFM
(T)	THERMOSTAT
	SQUARE TO ROUND TRANSITION
	PARALLEL BLADE DAMPER
	FIRE DAMPER FD (WALL) (FLOOR)
	AIRFOIL BLADE TURNING VANES
	AIR EXTRACTOR
ϕ	DIAMETER
¢	CFM (CUBIC FEET PER MINUTE)
RA	RETURN AIR
OSA	OUTSIDE AIR
CD	CONDENSATE DRAIN

104F28.EPS

Figure 28 ◆ HVAC legend.

3.3.6 Structural Member Symbols

Structural steel shapes have a system of identification by the use of symbols. Some examples of these symbols are shown in *Figure 29*. More detailed information about structural symbols is given later in this module.

3.3.7 Welding Symbols

Symbols for welding provide a means of conveying complete welding instructions from the designer to the welder. The symbols and their method of use are defined in the *American National Standard ANSI/AWS A2.4-79*, sponsored by the American Welding Society.

In *ANSI/AWS A2.4-79*, a distinction is made between the terms weld symbol and welding symbol. Weld symbols are used to indicate the type of welding to be performed. *Figure 30* shows some basic weld symbols.

A welding symbol is made up of as many as eight elements that are used together to provide exact welding instructions. *Figure 31* shows the standard elements of a welding symbol. As shown in *Figure 31*, the reference line is the key part of the welding symbol. All other elements are positioned with respect to this line. An arrow pointing to the location of the weld is positioned on one end of the reference line. When necessary, a tail is positioned on

the other end of the reference line. Other elements, such as the appropriate weld symbol used to show the type of weld, are placed on the reference line.

Information placed above the reference line indicates the weld is to be made on the other side of the joint from where the arrow points. Information placed below the reference line indicates the weld is to be made on the arrow side of the joint. Dimensions for the welds are drawn on the same side of the reference line as the weld symbol. Weld contour symbols are either flush, concave, or convex. The contour symbol is located next to the weld symbol. Finish symbols usually accompany the contour symbol. The weld groove angle is shown on the same side of the reference line as the weld symbol. The size (depth) is placed to the left of the weld symbol, and the root opening is shown inside the weld symbol.

3.4.0 Dimensioning

Dimensions given on drawings show actual sizes, distances, and heights of the objects and spaces being represented. Dimensions may be from outside to center, center to center, wall to wall, or outside to outside (*Figure 32*). In all cases, dimensions shown on drawings are given in full scale regardless of the fact that the plan shows an object or distance on a smaller scale.

BLUEPRINT READING FOR SURVEYORS

4.33

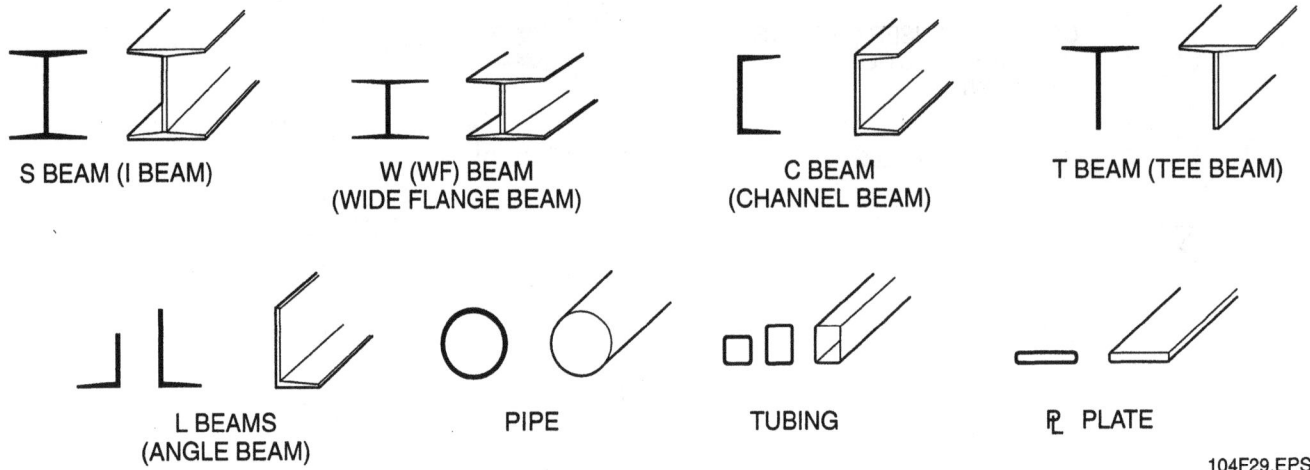

S BEAM (I BEAM)

W (WF) BEAM
(WIDE FLANGE BEAM)

C BEAM
(CHANNEL BEAM)

T BEAM (TEE BEAM)

L BEAMS
(ANGLE BEAM)

PIPE

TUBING

℄ PLATE

104F29.EPS

Figure 29 ◆ Structural member symbols.

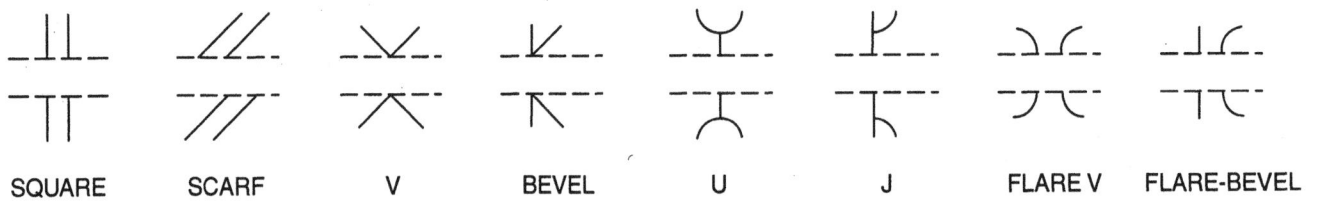

SQUARE   SCARF   V   BEVEL   U   J   FLARE V   FLARE-BEVEL

GROOVE WELD SYMBOLS

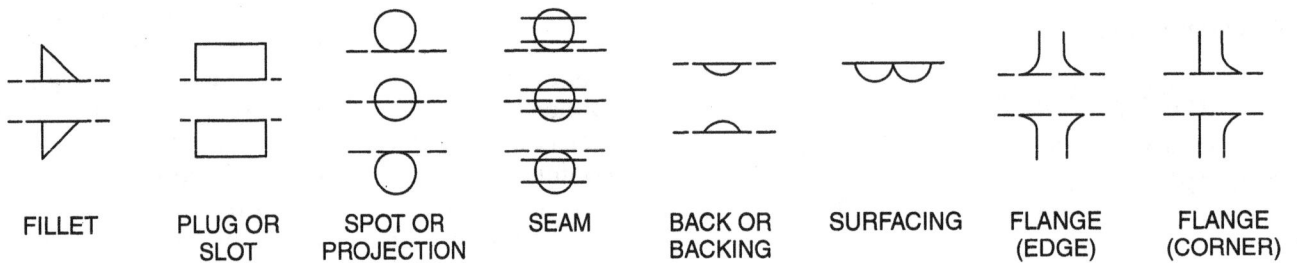

FILLET   PLUG OR
SLOT   SPOT OR
PROJECTION   SEAM   BACK OR
BACKING   SURFACING   FLANGE
(EDGE)   FLANGE
(CORNER)

OTHER WELD SYMBOLS

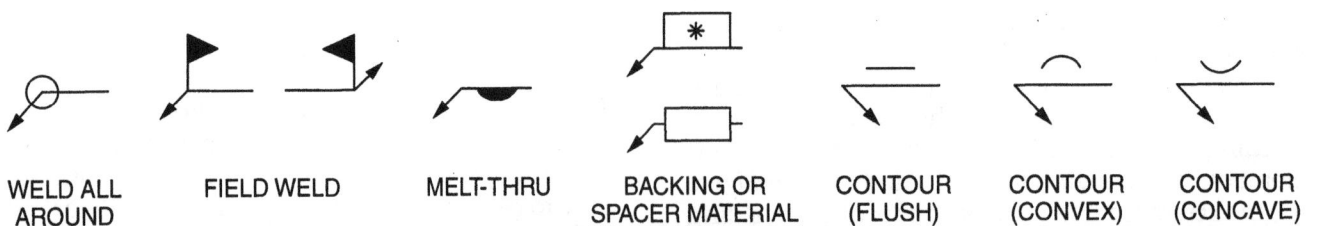

WELD ALL
AROUND   FIELD WELD   MELT-THRU   BACKING OR
SPACER MATERIAL   CONTOUR
(FLUSH)   CONTOUR
(CONVEX)   CONTOUR
(CONCAVE)

SUPPLEMENTARY SYMBOLS

104F30.EPS

Figure 30 ◆ Basic weld symbols.

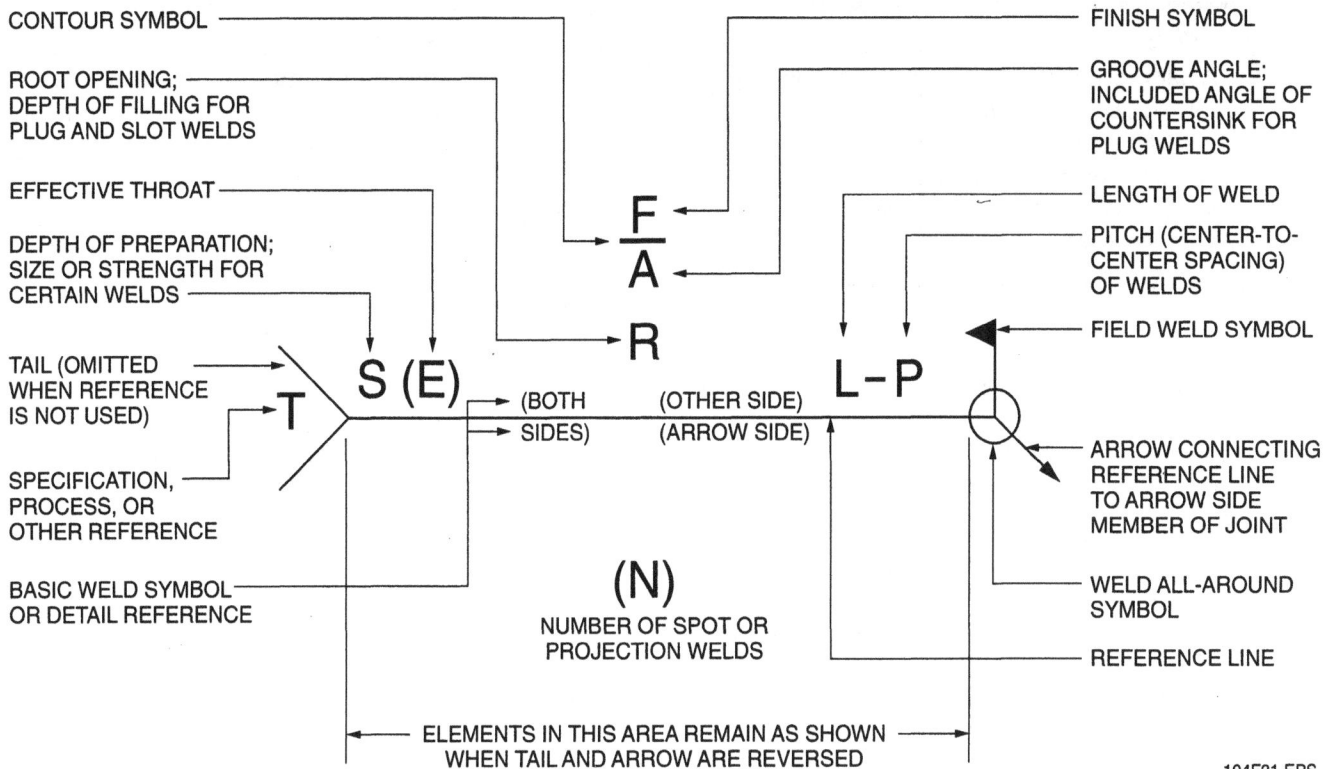

Figure 31 ◆ Elements of a welding symbol.

104F31.EPS

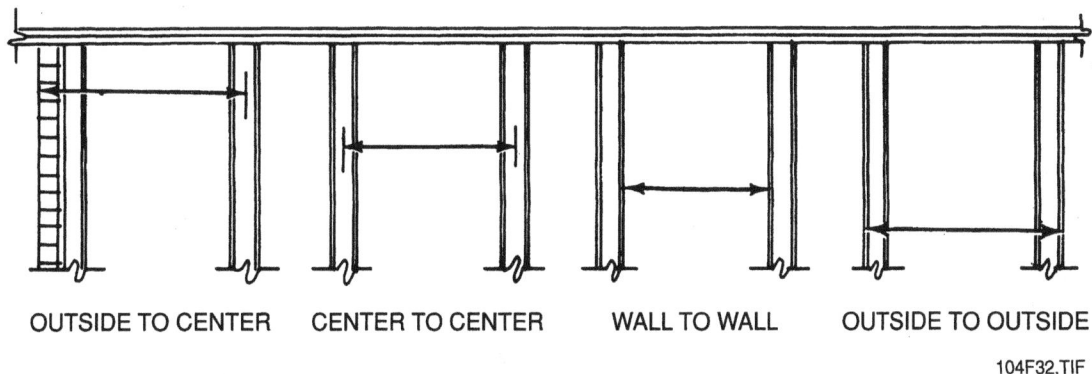

OUTSIDE TO CENTER   CENTER TO CENTER   WALL TO WALL   OUTSIDE TO OUTSIDE

104F32.TIF

Figure 32 ◆ Common methods of indicating dimensions on drawings.

Note that sectional and detail views may use a **nominal size** in labeling. Nominal sizes or dimensions are approximate or rough sizes by which lumber, block, etc., is commonly known and sold. For example, a nominal size 2 × 4 board is actually 1½" × 3½".

Sometimes it is necessary to determine dimensions not shown on a drawing by measuring them. Measurements are usually made on architectural drawings using an architect's scale rather than a standard ruler. Architect's scales are divided into feet and inches and usually consist of

several scales on one rule. Plan drawings are created using a specified scale. Inches or fractions of an inch on the drawing are used to represent feet in the actual measurement of a building. For example, in a plan drawn to ¼" scale, ¼" on the drawing represents 1' of the building. The scale of a drawing is usually shown directly below the drawing. The same scale may not be used for all the drawings that make up a complete set of plans. Methods for using the architect's rule to scale a drawing were covered earlier in the *Core Curriculum* module *Introduction to Blueprints*.

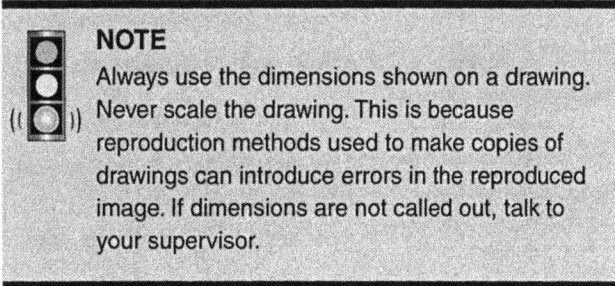

Some common practices used for dimensioning drawings are listed here. Keep in mind that these are not rules. The practices in your area may be different.

- Architectural dimension lines are unbroken lines, with the dimensions placed above and near the center of the line.
- Dimensions over one foot are shown in feet and inches (not decimals). Dimensions less than one foot are shown in inches only. The common exception to this rule is the center-to-center distances for standard construction, such as for framing 16" on center (OC) or 24" OC.
- Dimensions are placed to read from the right or from the bottom of the drawing.
- Overall building dimensions go to the outside of all other dimensions.
- Room sizes can be shown by stating width and length.
- Rooms are sometimes dimensioned from the center lines of partition walls, but wall-to-wall dimensions are more common.
- Window and door sizes are usually shown in window and door schedules.
- Dimensions that cannot be shown on the floor plan because of their size are placed at the end of leader lines.
- When stairs are dimensioned, the number of risers is placed on a line with an arrow indication in either an up or down direction.
- Architectural dimensions always refer to the actual size of the building, regardless of the scale of the drawings.

3.5.0 Abbreviations

Many written instructions are needed to complete a set of construction drawings. It is impossible to print out all such references, so a system of abbreviations is used. By using standard abbreviations, such as BRK for brick or CONC for concrete, the architect ensures that the drawings will be accurately interpreted. A brief list of some commonly used abbreviations is contained in *Appendix A*. Note that some architects and engineers may use different abbreviations for the same terms. Normally, the title sheet in a drawing set contains a list of abbreviations used in the drawings. For this reason, it is important to get a complete set of drawings and specifications, including the title sheet(s), so that you can better understand the exact abbreviations used. Some practices for using abbreviations on drawings are as follows:

- Most abbreviations are capitalized.
- Periods are used when abbreviations might look like a whole word.
- Abbreviations are the same whether they are singular or plural.
- Several terms have the same abbreviations and can only be identified from the context in which they are found.
- Many abbreviations are similar.

3.6.0 Architectural Terms Used in Drawings and Specifications

Many architectural terms are used in plans and specifications. Some commonly used terms are listed in *Appendix B*. You may already be familiar with may of the terms listed.

3.7.0 Orthographic Projection Drawings

An **orthographic projection drawing** is a type of drawing used to show the exact shape of an object. These drawings are typically used to show the specific details of parts or assemblies that need to be fabricated or manufactured. An orthographic projection drawing of an object is made by extending perpendicular (right-angle) lines from the object to create a series of projected plan views. The typical orthographic projection drawing uses three plan views to show the object. As shown in *Figure 33*, these are the front view, top view, and right-side view.

Each view shows the shape of the object as it is seen from a different direction or point of view. By mentally comparing the three or more views, a person can visualize what the object looks like.

The front view is considered the most important or primary view. It shows the surfaces visible on the front of the object. If the drawing is drawn to full scale, each line in the view is the same size as the corresponding front edges of the object. Obviously,

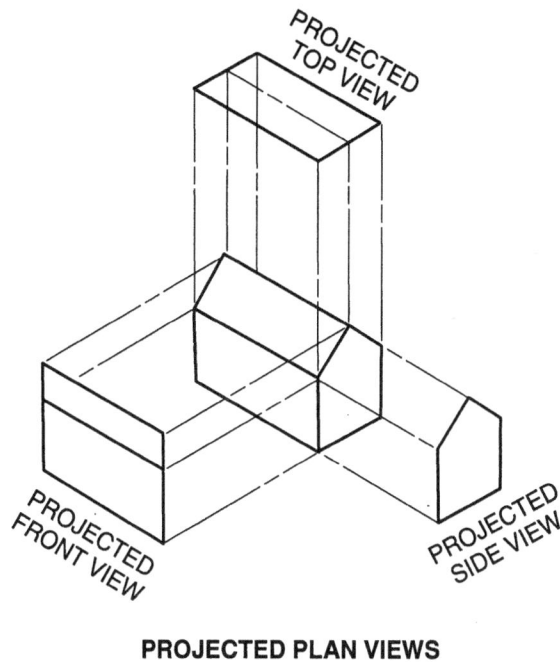

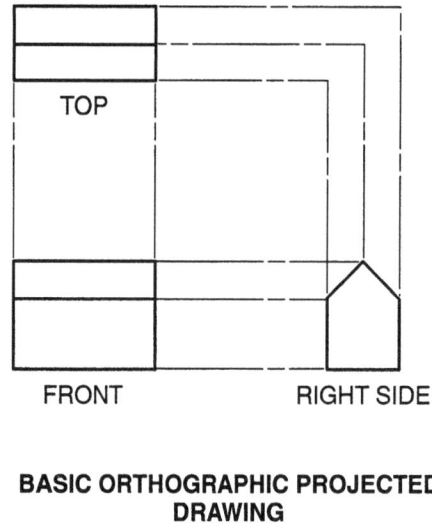

PROJECTED PLAN VIEWS

BASIC ORTHOGRAPHIC PROJECTED
DRAWING

104F33.EPS

Figure 33 ◆ Basic three-view orthographic projection drawing.

use of only a front view does not provide all the information needed to visualize an object. For example, the width of the object cannot be determined from the front view. A top view, along with the right-side view, are used to convey a more complete representation of the object. The position of the views is important. The top view is always shown above the front view so that their lengths line up. The right-side view is always placed to the right of the front view so that so that their heights line up. This arrangement of views is called a third angle projection.

Sometimes, important details of an object cannot be shown clearly by showing only the front, top, or right-side views. This may be true when showing an odd-shaped or complex object. In this case, additional views are displayed as needed to show the information. Left-side, back, and bottom views are added when they are needed to show important information. When drawn, these views are arranged in a particular order and labeled to eliminate confusion, as shown in *Figure 34*. As shown in the figure, dashed lines are used to represent features such as holes drilled in the object and edges that cannot be seen from a given point of view. As with other drawings, dimension lines are used to show the sizes and locations of features.

3.8.0 Isometric Drawings

Isometric drawings (ISOs) are three-dimensional drawings commonly used to show the shape of an object as it is viewed from one corner (*Figure 35*). An ISO drawing is commonly used in conjunction with a related orthographic projection drawing to provide all the information needed to fabricate or assemble an object.

Another common use for isometric drawings is to show the details necessary to fabricate and install a section of a piping system. It combines the plan and elevation views into one drawing. *Figure 36* shows a piping ISO drawing. ISO pipe drawings use the same symbols as any other piping drawing. To read an ISO pipe drawing, begin at the end of the pipe run nearest the source, and work in the direction of flow (the ISO has arrows to show direction of flow). Trace the line from the first fitting, and identify all the fittings and components by their symbols and abbreviations. Also, note the dimensions from one fitting or component to the other. The dimensions for valves and other components are shown on the ISO. The take-out for fittings and flanges must be calculated from the dimensions given on the ISO. The ISO also provides pipeline elevations for installation purposes.

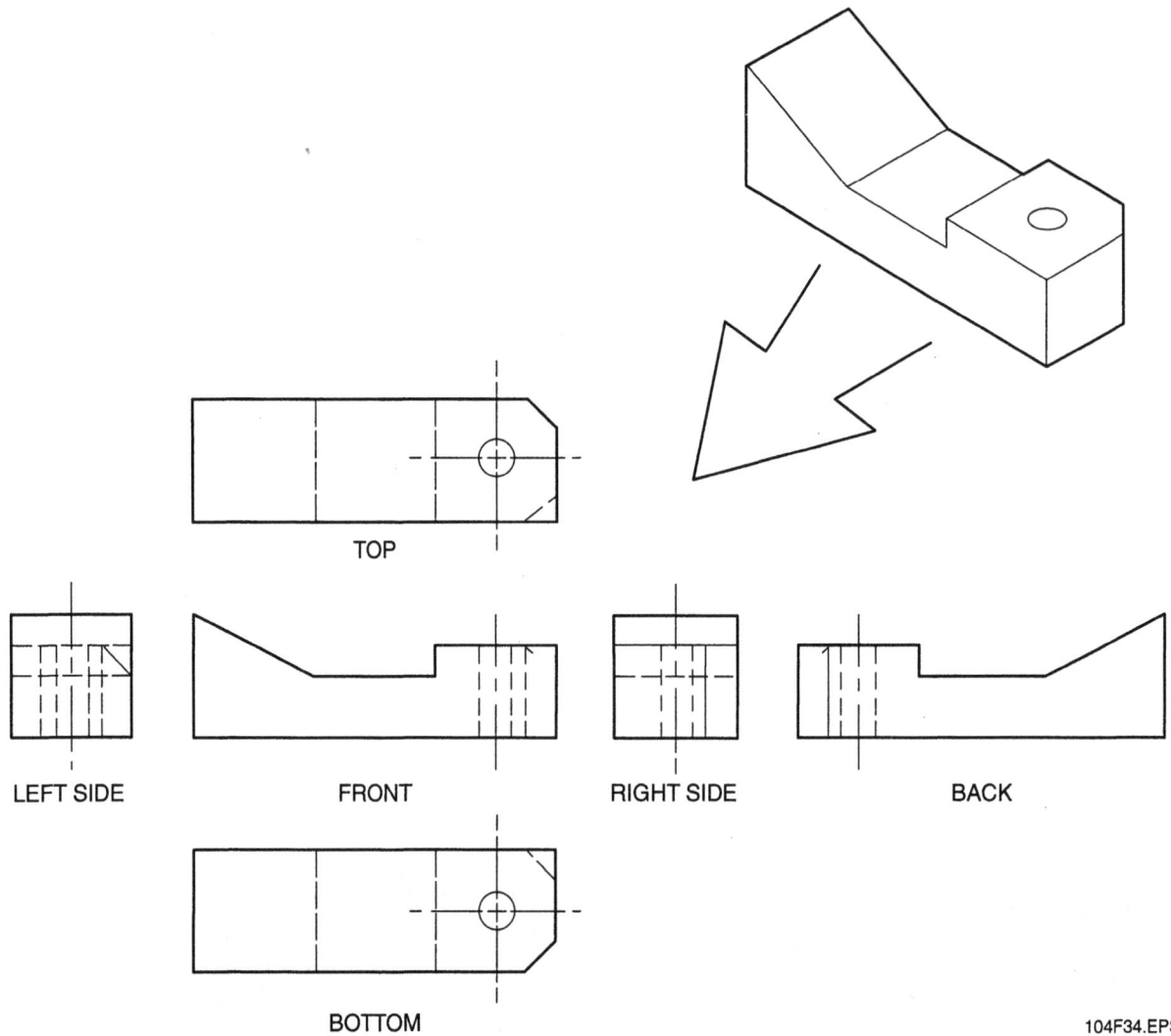

TOP

LEFT SIDE   FRONT   RIGHT SIDE   BACK

BOTTOM

104F34.EPS

Figure 34 ◆ Relationship of six views.

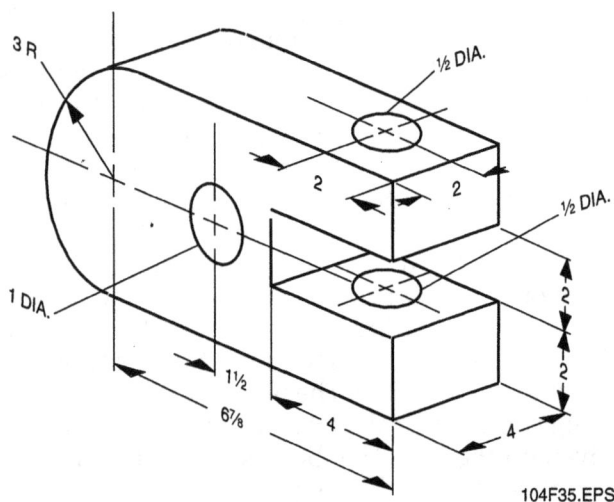

3 R
½ DIA.
½ DIA.
1 DIA.
2
2
2
2
2
1½
6⅞
4
4

104F35.EPS

Figure 35 ◆ Isometric drawing of an object.

4.0.0 ◆ STRUCTURAL STEEL AND CONCRETE CONSTRUCTION-RELATED DRAWINGS

Drawing sets for construction projects involving steel and/or concrete structures typically consist of all the same kinds of drawings described earlier in this module. This section introduces some additional types of detail, section, and shop drawings you may encounter in drawing sets for structures that incorporate structural steel and/or concrete. These include the following:

- Anchor bolt placing drawings
- Embed placing drawings
- **Erection drawings**
- Shop fabrication drawings
- Reinforcing steel placement drawings

4.38

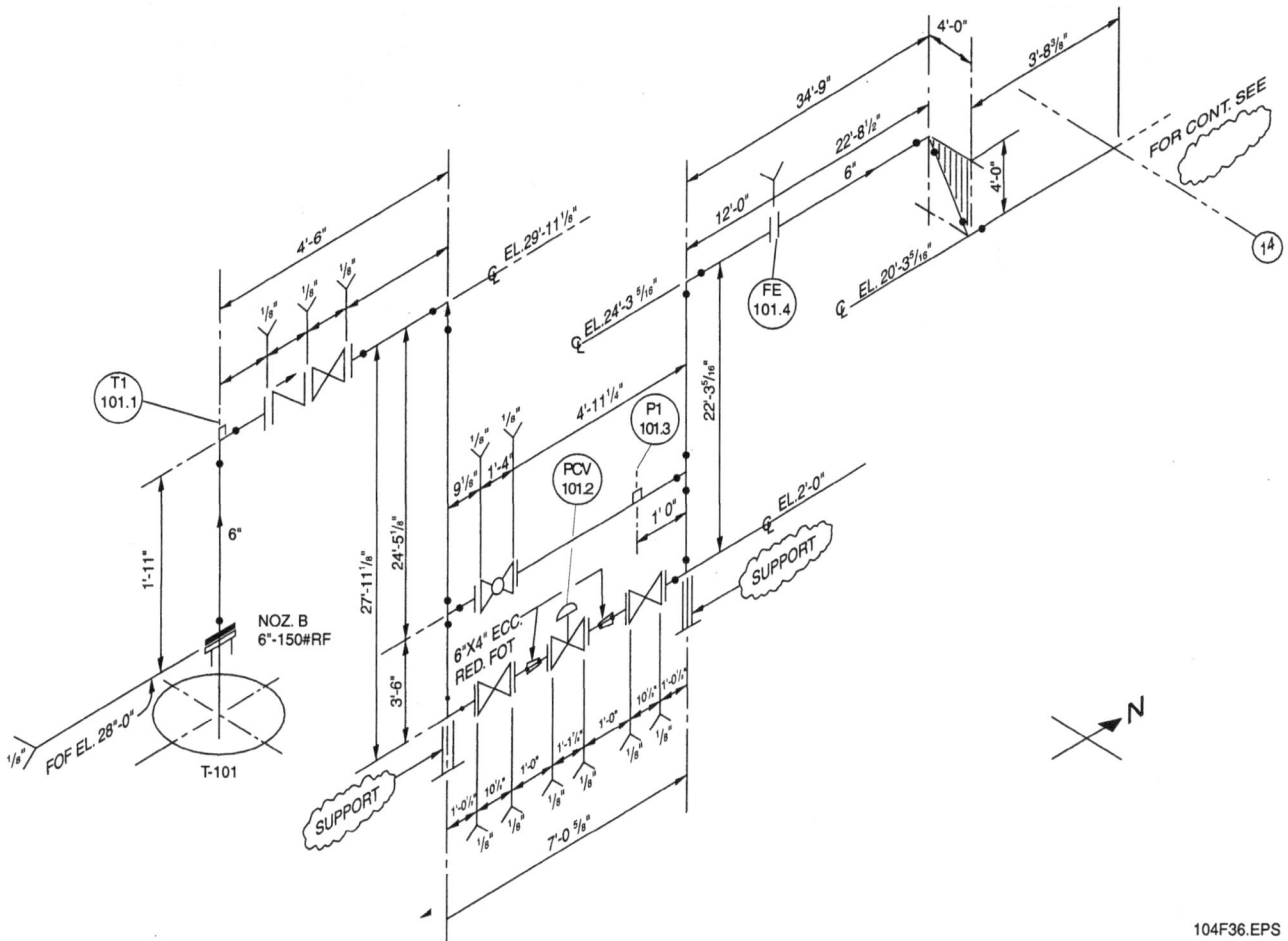

Figure 36 ◆ Isometric piping drawing.

104F36.EPS

4.1.0 Anchor Bolt Drawings

Anchor bolt placing drawings (*Figure 37*) are considered to be a form of shop drawing. Both the anchor bolts and placement drawings may be provided by a steel fabricator. Anchor bolt placing drawings are used when installing concrete and masonry support structures for steel columns, beams and other steel supports. They show the exact location and elevation of the anchor bolts for the steel installation. Anchor bolt drawings are typically used to show their placement relative to the various columns and beams in a structure.

Anchor bolts for columns are set in concrete grade beams, footings, concrete walls, and pilasters. The bolts must project a specified amount above the concrete or masonry to allow for grout to be placed under the base plate of the column for full bearing of the base plate to concrete. The grout space is achieved by stacking steel shims on which the base plate rests until grouted. The bolt spacing is such that it allows the bolts to be outside the column for more rigidity during erection. Anchor bolts for

beams are generally spaced to accommodate the hole gauge of the beam. Again, the beam is shimmed for a grout space for full bearing of the beam to concrete or masonry.

4.2.0 Embed Placing Drawings

Embed placing drawings show the proper location and elevation of plates embedded in concrete and masonry. Embed plates are used for beam bearing, joist bearing, column bearing, and other miscellaneous steel bearings in a structure. The drawings show the size of the embed plate and a piece mark for identification. *Figure 38* shows an example of an embed placement section drawing.

4.3.0 Erection Drawings

Erection drawings are plan drawings used mainly by ironworkers during a project to determine the location of the rigid frames and columns of a building. Most erection or framing drawings are nearly exact duplicates of the architectural floor or

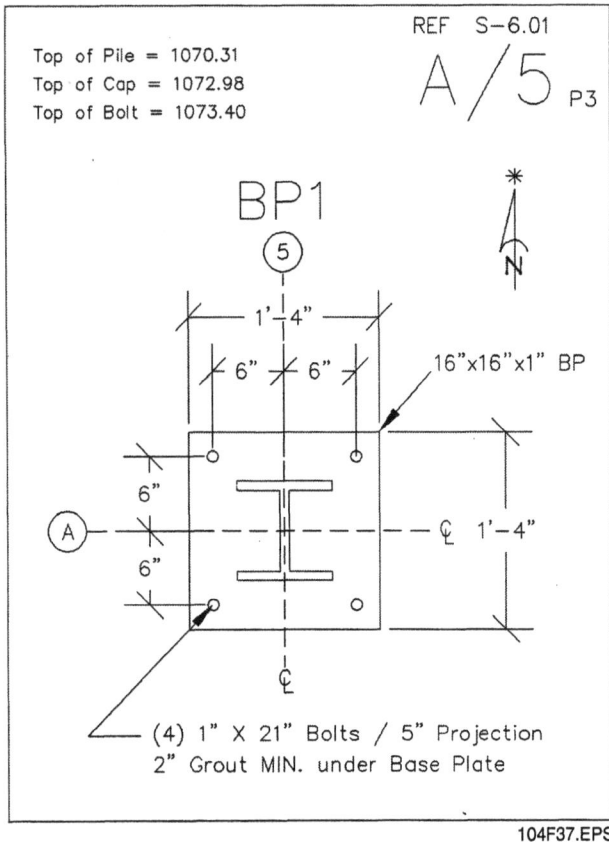

Top of Pile = 1070.31
Top of Cap = 1072.98
Top of Bolt = 1073.40

REF S-6.01

A/5 P3

BP1
(5)

1'-4"
6" 6"
16"x16"x1" BP

6"
(A)
6"
C 1'-4"

C

— (4) 1" X 21" Bolts / 5" Projection
2" Grout MIN. under Base Plate

104F37.EPS

Figure 37 ◆ Detailed anchor bolt placement drawing.

roof plan drawings, but without the information given for other trades. The beam sizes and column sizes are shown with a piece mark. Columns are shown with a symbol and a diagonal line indicating the size and piece mark on the line. If columns are stacked, the upper column is generally shown on top of the diagonal line, and the lower column is called out below the diagonal line. The sequence of erection is noted by the erection drawing numbering system. The sheets are generally numbered E1, E2, E3, and so on, or S1, S2, S3, and so on. The E1 sheet indicates it is the first sheet used.

Figure 39 shows an example of an erection (framing) plan for a second floor of a building. As shown in the figure, the plan is divided by grid lines. Each grid line is identified by an alphabetical letter or by a number. Generally, columns are located at the intersection points of grid lines. For example, if you follow grid line C to the intersection point of grid line 3 on the drawing, you find that a column identified as piece mark B8 is on the bottom, and a column identified as piece mark C10 is stacked on top. The bottom column, B8, is specified as a TS6 × 6 × ⅜. This means that this column is fabricated from 6" × 6" × ⅜" tube steel. The stack column, C10, is specified as TS4 × 4 × ¼. This means that it is fabricated from 4" × 4" × ¼" tube steel. You will also find that the beam that runs from grid C3 to grid D3 is a W21 × 62, with a piece

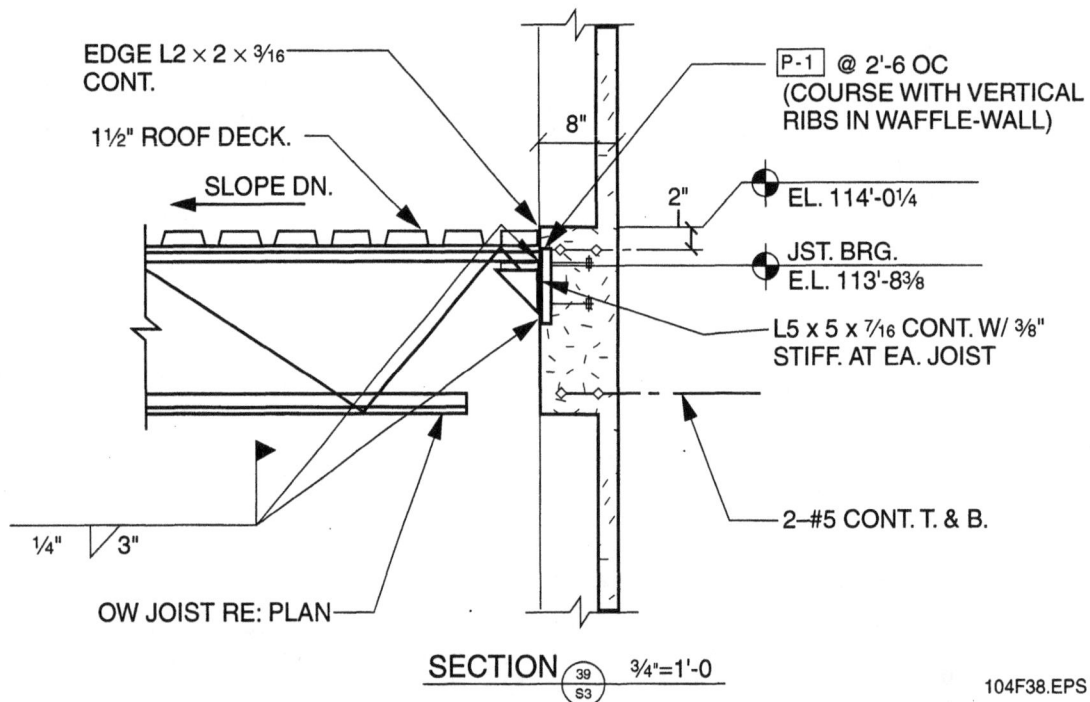

EDGE L2 × 2 × ³⁄₁₆ CONT.

1½" ROOF DECK.

SLOPE DN.

8"

P-1 @ 2'-6 OC
(COURSE WITH VERTICAL RIBS IN WAFFLE-WALL)

2"

EL. 114'-0¼

JST. BRG.
E.L. 113'-8⅜

L5 × 5 × ⁷⁄₁₆ CONT. W/ ⅜"
STIFF. AT EA. JOIST

¼" 3"

OW JOIST RE: PLAN

2–#5 CONT. T. & B.

SECTION 39
         S3   ¾"=1'-0

104F38.EPS

Figure 38. ◆ Embed placement section drawing.

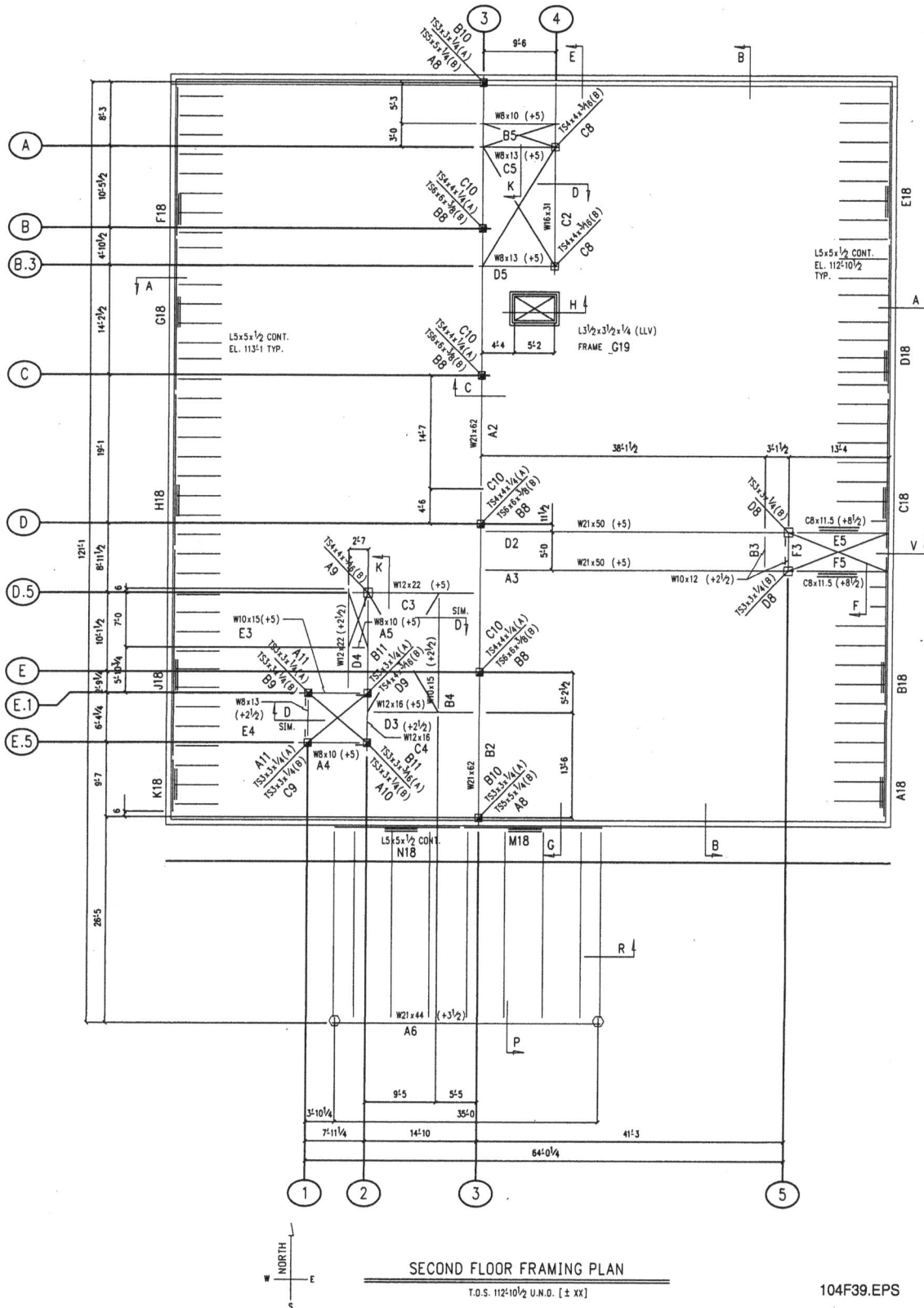

COLUMN LEGEND:
- ⊡ TS COLUMN BELOW (B)
- ◨ TS COLUMN BELOW (B) & ABOVE (A)
- ■ TS COLUMN ABOVE (A)
- ⬦ TS COLUMN THRU TOS ELEVATION

$\frac{1}{8}'' = 1'-0$

SECOND FLOOR FRAMING PLAN

T.O.S. 112'-10½ U.N.O. [± XX]

104F39.EPS

Figure 39 ◆ Erection drawing.

mark of A2. This means this is a W-type beam 21" high, and it weighs 62 pounds per foot of length. The method used to specify the different kinds of structural steel and other components is covered later in this module.

4.4.0 Shop Fabrication Drawings

Shop fabrication drawings are detailed drawings from which a fabricator builds a beam or other steel structural element. A fabricator is a person who provides detailed drawings for the fabrication of components and who fabricates them in a shop for later installation at the job site. The installer at the job site uses the drawings to identify the piece by shape and length. The drawings show the types of end connections and intermediate connections on beams and columns. They also show any shop-welded items that the installer may need for miscellaneous steel attachments, such as weld plates for brace angles. The shop fabrication drawing identifies all shop-installed items, welded or bolted on, in the shop fabrication process. Piece marks on erection drawings are referenced to the fabrication drawing by the first

number of the piece mark. Erection drawings are floor plan-type drawings that identify and show the location of all of the steel structural members. For example, a piece marked 5B1 is located on fabrication sheet 5. A piece marked 10C2 is on sheet 10. *Figure 40* shows an example of a typical fabrication drawing for a beam.

4.5.0 Reinforcing Steel Placement Drawings

Concrete by itself is considered strong in compression but weak in tension and shear. Whenever tension is present, the concrete is reinforced with steel. Reinforced concrete is a combination of steel and concrete that allows for the use of the best properties of each material.

As the name implies, reinforcing steel placement drawings show the types and location of reinforcing bars (rebars) to be used in reinforced concrete beams and other concrete structures. A structural engineer calculates the number, size, shape, and placement of reinforcing steel needed to meet all design requirements. The reinforcing steel is ordered to these specifications and placed

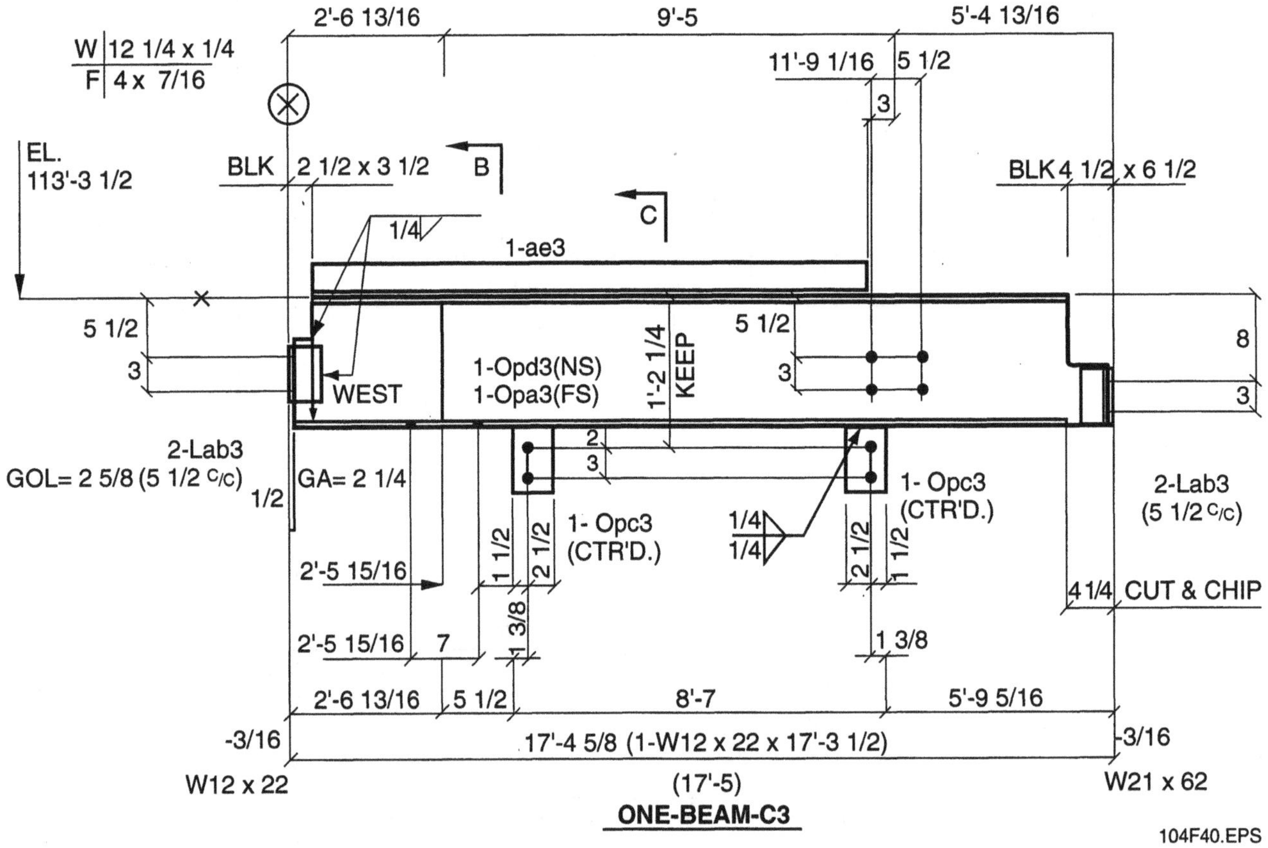

Figure 40 ◆ Beam fabrication drawing.

104F40.EPS

in the form for the concrete structure according to the related placement drawing. The reinforcing steel is then wired or held in position by appropriate specialized supports. *Figure 41* shows an example of a reinforcing steel placing drawing for a concrete beam.

A typical reinforced concrete beam like the one shown in *Figure 41* contains several horizontal reinforcing bars located near the top and bottom of the beam to resist tension at the middle of the span. However, when the beam passes over a support, the top of the beam is in tension; therefore, the bars are located near the top of the beam over the supports. The beam must also be designed to resist shear near its supports. This is done by the use of reinforcing steel stirrups (U-shaped bars) placed vertically.

4.6.0 Interpreting the Shapes and Types of Structural and Reinforcing Steel Shown on Drawings

The shape or type of structural steel is identified on structural drawings, erection drawings, and other drawings with a symbol or abbreviation. Size and dimensions are always given in a specified order. The specification format for designating the shape and size of common structural and reinforcing steel is described for each relevant shape in the following sections.

4.6.1 Plate

Plate is rolled metal of uniform thickness, having a thickness equal to or greater than ³⁄₁₆". Its identifying symbol is PL, and its thickness is given in inches. Its length and width are given in feet. For example, PL ¼" × 4' × 10'

4.6.2 Sheet Metal

Sheet metal is rolled metal of uniform thickness, having a thickness of less than ³⁄₁₆". Thickness is given in inches or in American Wire Gauge (AWG) number, and length and width are given in feet. For example: Sheet No. 12 AWG × 4' × 8'.

4.6.3 Bars

Bars are rolled to a variety of cross-section shapes and sizes (*Figure 42*). Standard bar shapes include the following:

- *Round* – Round bar is specified by BAR, followed by the bar diameter, a circle with a slash through it, and then the length.
- *Square* – Square bar is specified by BAR, followed by the face width, a square with a slash through it, and then the length.
- *Bar (flat)* – Flat bar is specified by BAR, followed by the wide dimension, the narrow dimension, and then the length.

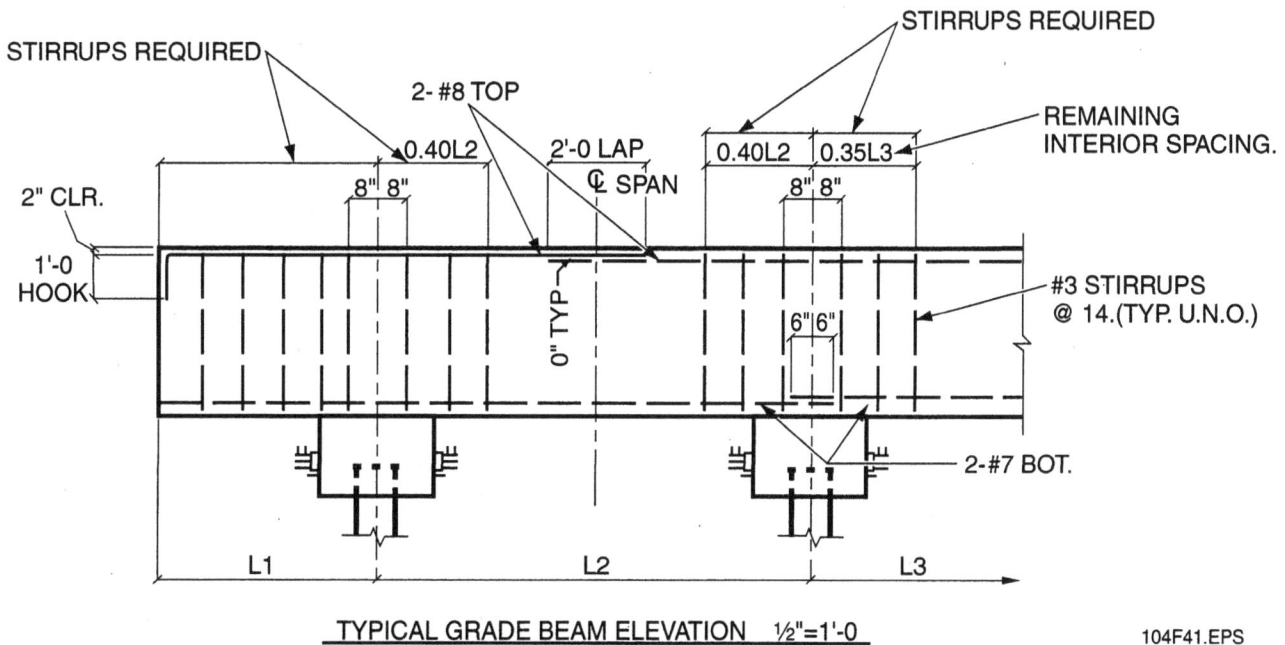

TYPICAL GRADE BEAM ELEVATION ½"=1'-0

104F41.EPS

Figure 41 ◆ Reinforcement steel placing drawing for a reinforced concrete beam.

SHAPE		SPECIFICATION FORMAT
ROUND	○	BAR ¾" ∅ × 8'-3"
SQUARE	□	BAR ¾" ⊡ × 8'-3"
BAR	▭	BAR 2" × ¼" × 8'-3"
Z-BAR	⌐	Z 2" × 2" × ½" × 8'-3"
HEXAGONAL	⬡	HEX ⅞" × 10'
OCTAGONAL	⯃	OCT 1" × 6'

104F42.EPS

Figure 42 ◆ Bar shapes and specification formats.

- *Z-bar* – Z-bar is specified by Z, followed by the flange width, the web depth, the flange and web thickness, and then the length. The flange and web are always the same thickness.
- *Hexagonal* – Hexagonal bar is specified by HEX, followed by the bar thickness (measured across the flats), and then the length.
- *Octagonal* – Octagonal bar is specified by OCT, followed by the bar thickness (measured across the flats), and then the length.

4.6.4 Angles

Angle is an L-shaped bar that can have either equal-sized or unequal-sized legs that are always at 90° angles to each other. *Figure 43* shows both types of angle bar stock and the specification formats for each type. The specification format represents the size and thickness of the angle bar stock legs in inches. Thus, L2 × 2 × ⅜ indicates a 2" × 2" angle of ⅜" thickness. For unequal leg stock, the longest leg is listed first, as shown in *Figure 43*.

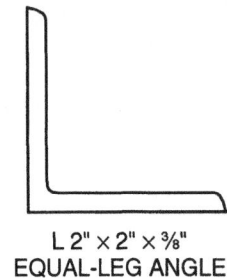

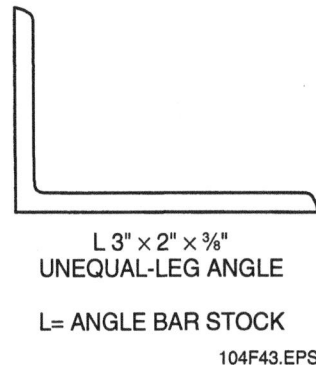

L 2" × 2" × ⅜"
EQUAL-LEG ANGLE

L 3" × 2" × ⅜"
UNEQUAL-LEG ANGLE

L= ANGLE BAR STOCK

104F43.EPS

Figure 43 ◆ Angle bar stock and specification formats.

4.6.5 Channels

Channels are U-shaped forms made of two flanges connected by a common web. The flanges extend from the same side of the web. The flanges can be of uniform thickness or tapered toward the outer edges. There are several channel variations, each having a unique specification format designation. *Figure 44* illustrates two different channels and their related American Standard Channels specification format designations. The channel designated as C8 × 11.5 represents a channel having a nominal depth of 8" and a weight of 11.5 pounds per foot of length. Similarly, the channel designated C8 × 18.75 has a nominal depth of 8" and a weight of 18.75 pounds per foot of length. As shown in the figure, the channel designated as C8 × 11.5 has a flange width of 2¼" and a web thickness of ¼", while the C8 × 18.75 channel has a flange width of 2¼" and a web thickness of ½". The specific dimensions for the beam depth, flange width, and web thickness for all common channel designations can be found in related tables of dimensions and properties that are provided in most structural engineering and architectural standards reference books.

In addition to channels specified in the American Standard Channel format, some channels are specified in the Miscellaneous Channels designation format. Miscellaneous channels are usually

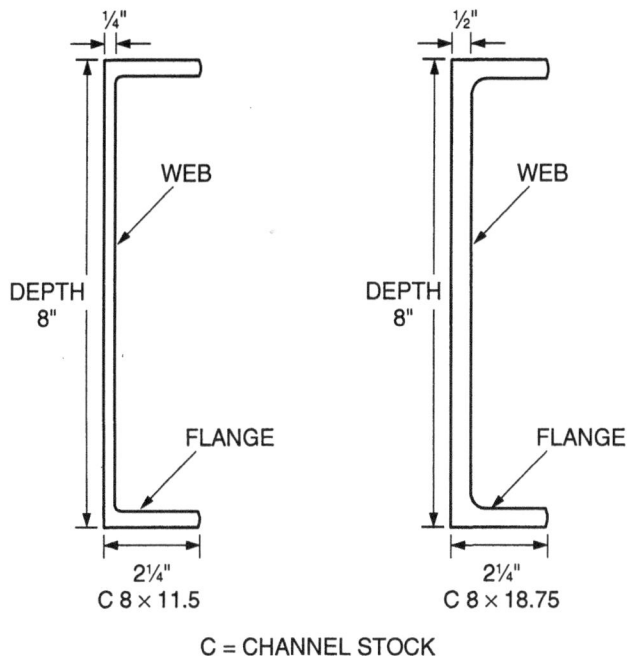

Figure 44 ◆ Channels and specification formats.

lighter channels than American Standard channels. The only difference in the specification format is that the format for the miscellaneous channels starts with the letters MC instead of C, as used with the American Standard channels.

4.6.6 Beams and Shapes from Beams

Beams (*Figure 45*) are made in I-shaped, H-shaped, and T-shaped cross sections and are made with flat or tapered flanges. In addition to different flange widths and web depths, the thicknesses of the flanges and webs vary with beam sizes. Also, beam weight (pounds per linear foot) for a beam of a given dimension can be increased by adding thickness to the web and flanges with very little change in beam width or depth. The specification designator describes the beam type, its nominal depth in inches, and the weight of the beam per foot of length. An example specification designator for a beam is $S24 \times 120$, where S24 represents the beam type (S) and nominal depth in inches (24). The 120 is the weight in pounds per foot of length. The specific dimensions for the depth, web thickness, and flange thickness for the various designations of beams are given in related tables of dimensions and properties in most structural engineering and architectural standards books.

Beams are identified with a standard set of symbols. The I-beam and H-beam letter symbols include:

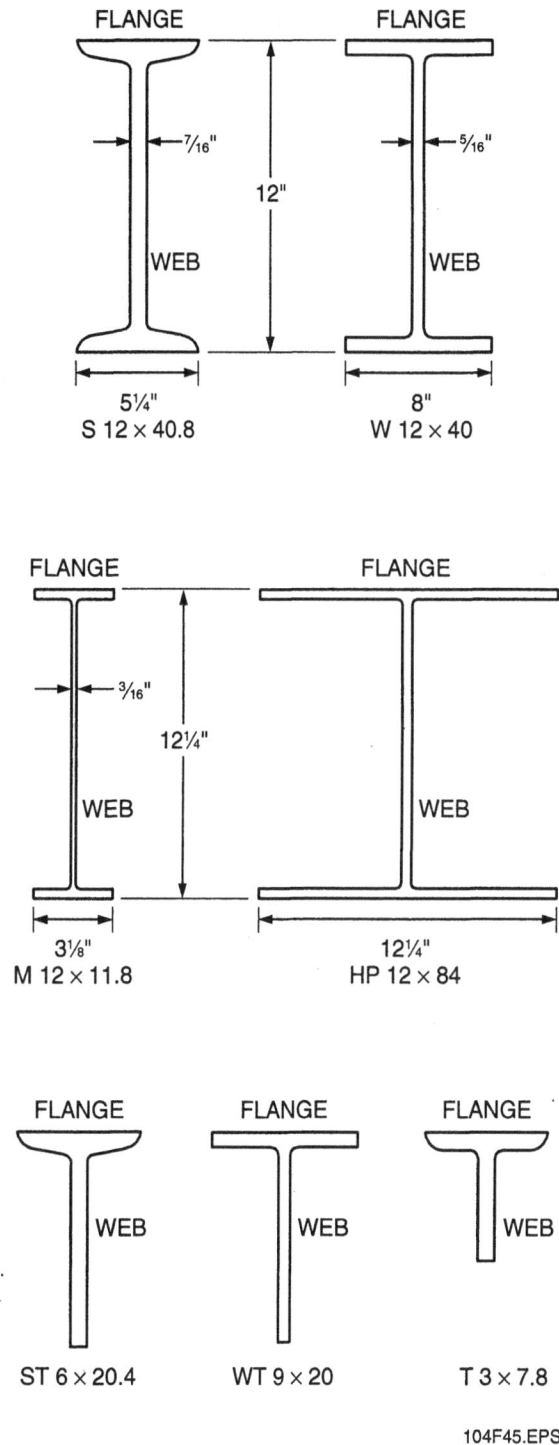

Figure 45 ◆ Structural beam and T-beam shapes.

- *S (American Standard)* – I-shaped beams with tapered flanges
- *W (Wide flange)* – Wider flanges than S beams with thinner webs and nontapered flanges
- *M* – An additional classification for beams other than S or W beams
- *HP* – H-shaped beams with nontapered flanges

T-beams include structural tees and tee shapes. Structural tees are made by cutting S, W, and M beams down the center, usually by shearing. The conventional symbol for a structural tee is the original beam letter followed by a T. For example, a tee cut from an S8 × 18.4 beam would be identified as ST4 × 9.2.

The symbol for tee shapes is a T, without any other letter. The specification gives the nominal depth, flange width, thickness, and length. An example specification is T3 × 7.8. T-shapes are short pieces, commonly used as connections or supports.

4.6.7 Pipe

Pipe size is specified as a nominal size that approximates its inside diameter (ID) up through 12". Pipe 14" and larger is specified by the actual OD (outside diameter). Pipe weight (determined by the wall thickness) is specified by a schedule number or strength. Schedule numbers are 10, 20, 30, 40, 60, 80, 100, 120, 140, and 160. Strengths are specified as standard (STD), extra strong (XS), and extra-extra strong (XXS). Standard falls between Schedules 30 and 40. Extra strong falls between Schedules 60 and 80, and extra-extra strong is Schedule 160 or higher. *Table 1* lists standard pipe sizes and schedules with nominal wall thicknesses.

Because the OD of a given pipe size is always the same, the ID becomes smaller as the pipe wall thickness increases with higher schedule pipe.

4.6.8 Reinforcing Bars

Reinforcing bars are used for concrete reinforcement. They are available in several grades. These grades vary in yield strength, ultimate strength, percentage of elongation, bend-test requirements, and chemical composition. Reinforcing bars can be coated with different compounds, such as epoxy, for use in concrete where corrosion could be a problem. The American Society for Testing and Materials International (ASTM) has established standard specifications for reinforcing bars. These grades appear on bar-bundle tags, in color coding, in rolled-on markings on the bars, and/or on bills of materials. The specifications are as follows:

Table 1 Standard Pipe Sizes and Schedules with Nominal Wall Thicknesses

Nominal Pipe Size	Outside diameter	Sched. 10	Sched. 20	Sched. 30	STD	Sched. 40	Sched. 60	XS	Sched. 80	Sched. 100	Sched. 120	Sched. 140	Sched. 160	XXS
					NOMINAL WALL THICKNESS									
1/8	0.405	–	–	–	0.068	0.068	–	0.095	0.095	–	–	–	–	–
1/4	0.540	–	–	–	0.088	0.088	–	0.119	0.119	–	–	–	–	–
3/8	0.675	–	–	–	0.091	0.091	–	0.126	0.126	–	–	–	–	–
1/2	0.840	–	–	–	0.109	0.088	–	0.147	0.147	–	–	–	0.188	0.294
3/4	1.050	–	–	–	0.113	0.113	–	0.154	0.154	–	–	–	0.219	0.308
1	1.315	–	–	–	0.133	0.133	–	0.179	0.179	–	–	–	0.250	0.358
1-1/4	1.660	–	–	–	0.140	0.140	–	0.191	0.191	–	–	–	0.250	0.382
1-1/2	1.900	–	–	–	0.145	0.145	–	0.200	0.200	–	–	–	0.281	0.400
2	2.375	–	–	–	0.154	0.154	–	0.218	0.218	–	–	–	0.344	0.436
2-1/2	2.875	–	–	–	0.203	0.203	–	0.276	0.276	–	–	–	0.375	0.552
3	3.5	–	–	–	0.216	0.216	–	0.300	0.300	–	–	–	0.438	0.600
3-1/2	2.875	–	–	–	0.226	0.226	–	0.318	0.318	–	–	–	–	–
4	4.5	–	–	–	0.237	0.237	–	0.337	0.337	–	0.438	–	0.531	0.674
5	5.563	–	–	–	0.258	0.258	–	0.375	0.375	–	0.500	–	0.625	0.750
6	6.625	–	–	–	0.280	0.280	–	0.432	0.432	–	0.562	–	0.719	0.864
8	8.625	–	0.250	0.277	0.322	0.322	0.406	0.500	0.500	0.594	0.719	0.812	0.906	0.875
10	10.75	–	0.250	0.307	0.365	0.365	0.500	0.500	0.594	0.719	0.844	1.000	1.125	1.000
12	12.75	–	0.250	0.330	0.375	0.406	0.562	0.500	0.688	0.844	1.000	1.125	1.312	1.000
14 OD	14.0	0.250	0.312	0.375	0.375	0.438	0.594	0.500	0.750	0.938	1.094	1.250	1.406	–
16 OD	16.0	0.250	0.312	0.375	0.375	0.500	0.656	0.500	0.844	1.031	1.219	1.438	1.594	–
18 OD	18.0	0.250	0.312	0.438	0.375	0.562	0.750	0.500	0.988	1.156	1.375	1.562	1.781	–
20 OD	20.0	0.250	0.375	0.375	0.375	0.594	0.812	0.500	1.031	1.281	1.500	1.750	1.969	–
22 OD	22.0	0.250	0.375	0.500	0.375	–	0.875	0.500	1.125	1.375	1.625	1.875	2.125	–
24 OD	24.0	0.250	0.375	0.562	0.375	0.688	0.969	0.500	1.218	1.531	1.812	2.062	2.344	–
22 OD	26.0	0.312	0.500	–	0.375	–	–	0.500	–	–	–	–	–	–
28 OD	28.0	0.312	0.500	0.625	0.375	–	–	0.500	–	–	–	–	–	–
30 OD	30.0	0.312	0.500	0.625	0.375	–	–	0.500	–	–	–	–	–	–
32 OD	32.0	0.312	0.500	0.625	0.375	0.688	–	0.500	–	–	–	–	–	–
34 OD	34.0	0.312	0.500	0.625	0.375	0.688	–	0.500	–	–	–	–	–	–
36 OD	36.0	0.312	0.500	0.625	0.375	0.750	–	0.500	–	–	–	–	–	–
42 OD	42.0	–	–	–	0.375	–	–	0.500						

104T01.EPS

- *A615, Standard Specification for Deformed and Plain Billet-Steel Bars for Concrete Reinforcement*
- *A616, Standard Specification for Rail-Steel Deformed Bars for Concrete Reinforcement*
- *A617, Standard Specification for Axle-Steel Deformed Bars for Concrete Reinforcement*
- *A706, Standard Specification for Low-Alloy Steel Deformed Bars for Concrete Reinforcement*

The standard configuration for reinforcing bars is the deformed bar. Different patterns may be impressed upon the bars depending on which mill manufactured them, but all are rolled to conform to ASTM specifications. The purpose of the deformation is to improve the bond between the concrete and the bar and prevent the bar from moving in the concrete. Plain bars are smooth and round without deformations on them and are used for special purposes, such as for dowels at expansion joints where the bars must slide in a sleeve, for expansion and contraction joints in highway pavement, and for column spirals. Deformed bars are designated by a number in eleven standard sizes (metric or inch-pound), as shown in *Table 2*. The number denotes the approximate diameter of the bar in the decimal equivalent of eighths of an inch or millimeters (mm). For example, a #5 bar has an approximate diameter of ⅝" (0.625). The nominal dimension of a deformed bar (nominal does not include the deformation) is equivalent to that of a plain bar having the same weight per foot.

As shown in *Figure 46*, bar identification is accomplished by ASTM specifications, which require that each bar manufacturer roll the following information onto the bar:

- A letter or symbol to indicate the manufacturer's mill
- A number corresponding to the size number of the bar (*Table 2*)
- A symbol or marking to indicate the type of steel (*Table 3*)
- A marking to designate the grade (*Table 4*)

The grade represents the minimum yield (tension strength), measured in **kips** per square inch (ksi) or megapascals (MPa), that the type of steel used will withstand before it permanently stretches (elongates) and will not return to its original length. Reinforcing bars, or rebars, are normally supplied from the mill bundled in 60' lengths, then are cut in the field to the required length. Straight bars are cut to length from longer stock in a fabricator's shop. Bent bars are cut to length from straight stock and are sent to be bent by a fabricator as specified.

There are many types of bends. The most common types (*Figure 47*) have been standardized by the American Concrete Institute. The bend type numbers are shown in drawing schedules or on bar lists. All of the bend types have a letter assigned to each dimension of the bar. When all of the dimensions are used on a bar list, the exact shape will be produced.

Table 2 ASTM Standard Metric and Inch-Pound Reinforcing Bars

Bar Size		Nominal Characteristics					
		Diameter		Cross-Sectional Area		Weight	
Metric	[in.-lb]	mm	[in.]	mm	[in.]	kg/m	[lbs/ft.]
#10	[#3]	9.5	[0.375]	71	[0.11]	0.560	[0.376]
#13	[#4]	12.7	[0.500]	129	[0.20]	0.944	[0.668]
#16	[#5]	15.9	[0.625]	199	[0.31]	1.552	[1.043]
#19	[#6]	19.1	[0.750]	284	[0.44]	2.235	[1.502]
#22	[#7]	22.2	[0.875]	387	[0.60]	3.042	[2.044]
#25	[#8]	25.4	[1.000]	510	[0.79]	3.973	[2.670]
#29	[#9]	28.7	[1.128]	645	[1.00]	5.060	[3.400]
#32	[#10]	32.3	[1.270]	819	[1.27]	6.404	[4.303]
#36	[#11]	35.8	[1.410]	1006	[1.56]	7.907	[5.313]
#43	[#14]	43.0	[1.693]	1452	[2.25]	11.380	[7.650]
#57	[#18]	57.3	[2.257]	2581	[4.00]	20.240	[13.600]

104T02.EPS

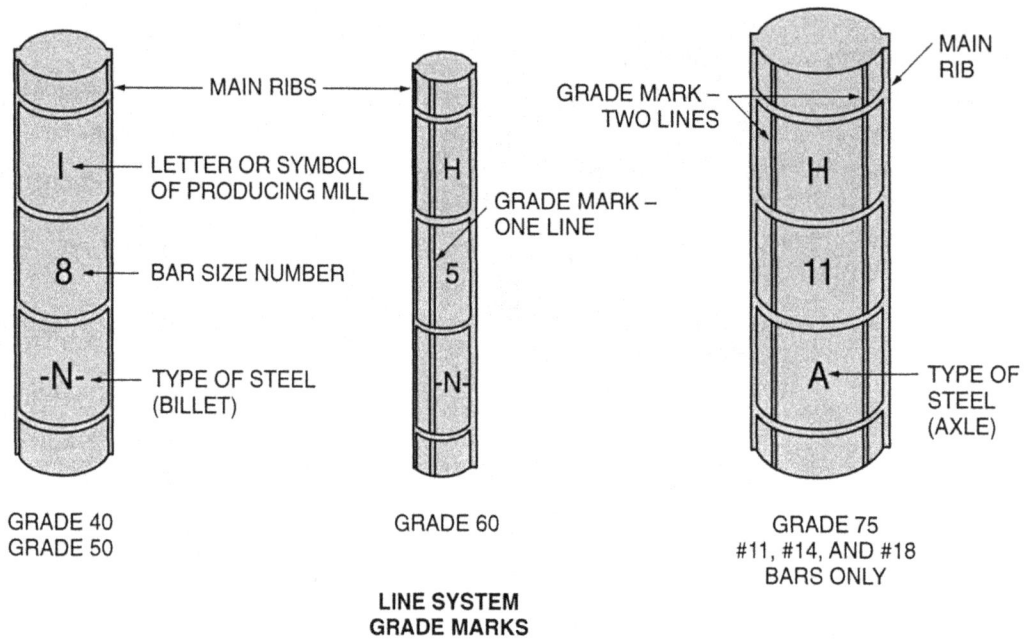

GRADE 40
GRADE 50

GRADE 60

GRADE 75
#11, #14, AND #18
BARS ONLY

**LINE SYSTEM
GRADE MARKS**

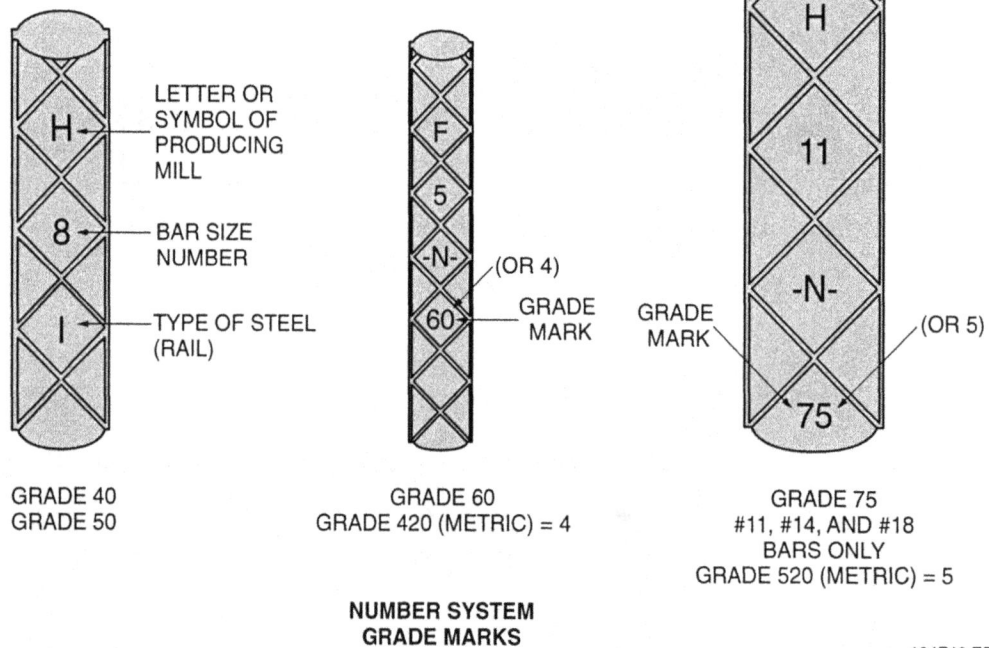

GRADE 40
GRADE 50

GRADE 60
GRADE 420 (METRIC) = 4

GRADE 75
#11, #14, AND #18
BARS ONLY
GRADE 520 (METRIC) = 5

**NUMBER SYSTEM
GRADE MARKS**

104F46.EPS

Figure 46 ◆ Reinforcement bar identification.

Table 3 Reinforcement Bar Steel Types

Symbol/Marking	Type of Steel
A	Axle (ASTM A617)
S or N	Billet (ASTM A615)
I or IR	Rail (ASTM A616)
W	Low-alloy (ASTM A706)

Table 4 Reinforcement Bar Grades

Grade	Identification	Minimum Yield Strength
40 and 50	None	40,000 to 50,000 psi (40 to 50 ksi)
60	One line or the number 60	60,000 psi (60 ksi)
75	Two lines or the number 75	75,000 psi (75 ksi)
420	The number 4	420 MPa (60,000 psi or 60 ksi)
520	The number 5	520 MPa (75,000 psi or 75 ksi)

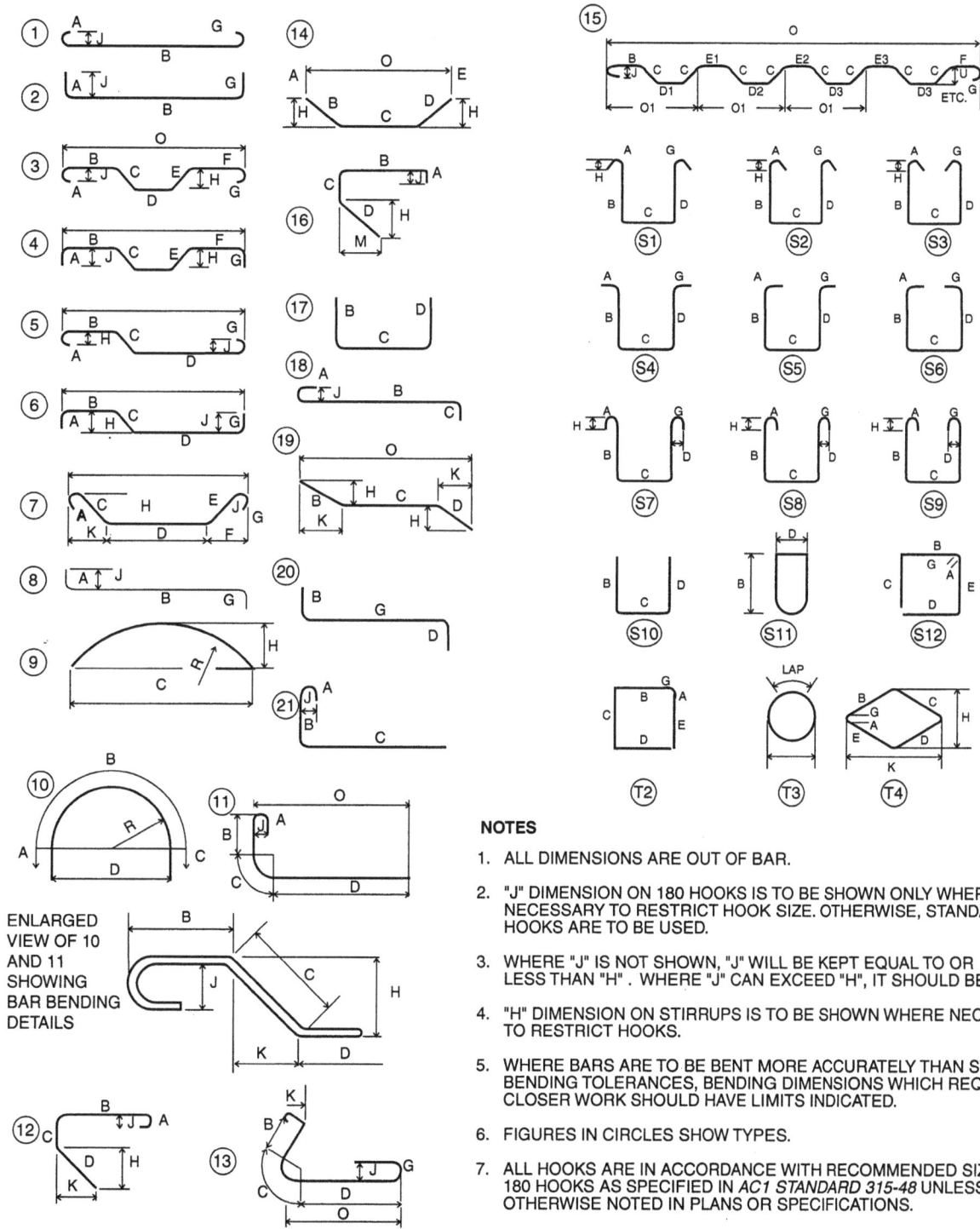

NOTES

1. ALL DIMENSIONS ARE OUT OF BAR.

2. "J" DIMENSION ON 180 HOOKS IS TO BE SHOWN ONLY WHERE NECESSARY TO RESTRICT HOOK SIZE. OTHERWISE, STANDARD HOOKS ARE TO BE USED.

3. WHERE "J" IS NOT SHOWN, "J" WILL BE KEPT EQUAL TO OR LESS THAN "H". WHERE "J" CAN EXCEED "H", IT SHOULD BE SHOWN.

4. "H" DIMENSION ON STIRRUPS IS TO BE SHOWN WHERE NECESSARY TO RESTRICT HOOKS.

5. WHERE BARS ARE TO BE BENT MORE ACCURATELY THAN STANDARD BENDING TOLERANCES, BENDING DIMENSIONS WHICH REQUIRE CLOSER WORK SHOULD HAVE LIMITS INDICATED.

6. FIGURES IN CIRCLES SHOW TYPES.

7. ALL HOOKS ARE IN ACCORDANCE WITH RECOMMENDED SIZES FOR 180 HOOKS AS SPECIFIED IN *AC1 STANDARD 315-48* UNLESS OTHERWISE NOTED IN PLANS OR SPECIFICATIONS.

104F47.EPS

Figure 47 ◆ Common types of rebar bends.

4.6.9 Tubing

Tubing is manufactured in square, rectangular, and round cross sections. Round tubing can be distinguished from pipe by its dimensions. Standard round tubing is measured by a nominal OD. The cross sections of square and rectangular tubing have even outside dimensions and slightly rounded corners. *Figure 48* shows standard tubing shapes and specification formats.

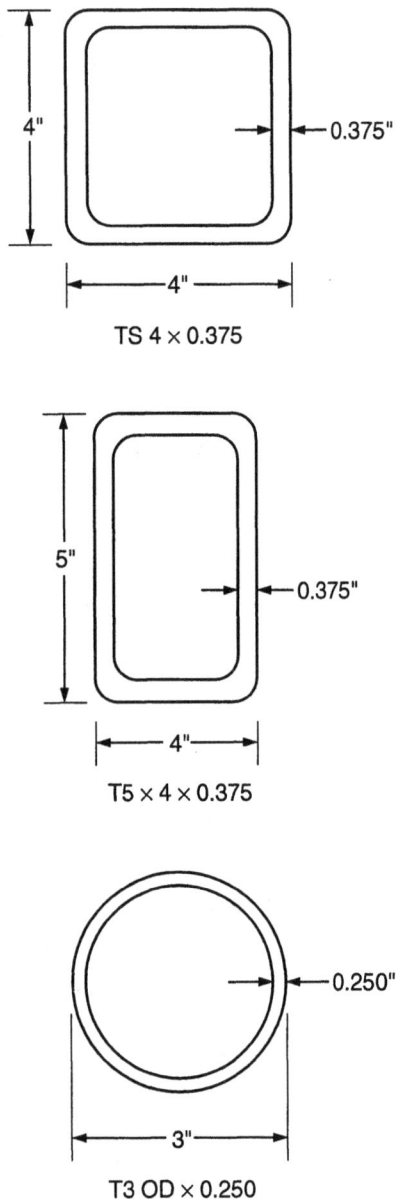

TS 4 × 0.375

T5 × 4 × 0.375

T3 OD × 0.250

104F48.EPS

Figure 48 ◆ Standard tubing shapes and specification formats.

5.0.0 ◆ GUIDELINES FOR READING A DRAWING SET

The following general procedure is suggested as a method of reading a set of drawings for maximum understanding:

Step 1 Acquire a complete set of drawings and specifications, including the title sheet(s), so that you can better understand the abbreviations and symbols used throughout the drawings.

Step 2 Read the title block. The title block tells you what the drawing is about. Take note of the critical information such as the scale, date of last revision, drawing number, and architect or engineer. After you use a sheet from a set of drawings, be sure to refold the sheet with the title block facing up.

Step 3 Find the north arrow. Always orient yourself to the structure. Knowing where north is enables you to more accurately describe the location of walls and other parts of the building.

Step 4 Always be aware that the drawings work together as a group. The reason the architect or engineer draws plans, elevations, and sections is that drawings require more than one type of view to communicate the whole project. Learn how to use more than one drawing when necessary to find the information you need.

Step 5 Check the list of drawings in your set. Note the sequence of the various plans. Some drawings have an index on the front cover. Notice that the prints in the set are of several categories:

- Architectural
- Structural
- Mechanical
- Electrical
- Plumbing

Step 6 Study the plot/site plan to determine the location of the building to be constructed, as well as the various utilities, roadways, and any easements. Note the various elevations and contours.

Step 7 Check the floor plan for the orientation of the building. Observe the locations and features of entries, corridors, offsets, and any special features.

Step 8 Check the foundation plan for the sizes and types of footings, reinforcing steel, and loadbearing substructures.

Step 9 Check the floor and wall construction and other details relating to exterior and interior walls.

Step 10 Study the plumbing, mechanical, electrical, and structural plans for features that extend for more than one floor, such as plumbing and vents, stairways, elevator shafts, and heating and cooling ductwork.

Step 11 Check the notes on various pages, and compare the specifications against the construction details.

Step 12 Thumb through the sheets of drawings until you are familiar with all the plans and structural details.

Step 13 Recognize applicable symbols and their relative locations in the plans. Note any special construction details or variations that will affect the carpentry work.

6.0.0 ◆ SPECIFICATIONS

Specifications, commonly called specs, are written instructions provided by architectural and engineering firms to the general contractor and, consequently, to the subcontractors. Specifications are just as important as the drawings in a set of plans. They furnish what the drawings cannot in that they define the quality of work to be done and the materials to be used. Specifications serve several important purposes:

• Clarify information that cannot be shown on the drawings

• Identify work standards, types of materials to be used, and the responsibility of various parties to the contract

• Provide information on details of construction

• Serve as a guide for contractors bidding on the construction job

• Serve as a standard of quality for materials and workmanship

• Serve as a guide for compliance with building codes and zoning ordinances

• Provide the basis of agreement between the owner, architect, and contractors in settling any disputes

Specifications are legal documents. When there is a difference between the drawings and the specifications, the specifications normally take legal precedence over the working drawings. However, the plans are often more specific to the job than the specifications. Therefore, notes on the plans may be considered by the architect/owner to be the true intent. You must be very careful to watch for discrepancies between the plans and specifications and report them to your supervisor immediately.

6.1.0 Organization and Types of Specifications

Specifications consist of various elements that may differ somewhat for particular construction jobs. For small projects, they may be simple; for large projects, they may be complex. Basically, two types of information are contained in a set of specifications: special and general conditions, and technical aspects of construction.

6.1.1 Special and General Conditions

Special and general conditions cover the non-technical aspects of the contractual agreements. Special conditions cover topics such as safety and temporary construction. General conditions cover the following points of information:

• Contract terms

• Responsibilities for examining the construction site

• Types and limits of insurance

• Permits and payments of fees

• Use and installation of utilities

• Supervision of construction

• Other pertinent items

The general conditions section is the area of the construction contract where misunderstandings often occur. Therefore, these conditions are usually much more explicit on large, complicated construction projects. An example of a typical residential material specification is shown in *Figure 49*.

> **NOTE**
> Residential specifications often do not spell out general conditions and are basically material specifications only.

Form RD 1924-2
(Rev. 7-99)

UNITED STATES DEPARTMENT OF AGRICULTURE
U.S. DEPARTMENT OF HOUSING AND URBAN DEVELOPMENT
FEDERAL HOUSING ADMINISTRATION
U.S. DEPARTMENT OF VETERANS AFFAIRS

FORM APPROVED
OMB NO. 0575-0042

☐ **Proposed Construction**

☐ **Under Construction**

DESCRIPTION OF MATERIALS

No. ________________ (To be inserted by Agency)

Property address ________________ *City* ________________ *State* Oklahoma

Mortgagor or Sponsor ________________ (Name) ________________ (Address)

Contractor or Builder ________________ (Name) ________________ (Address)

INSTRUCTIONS

1. For additional information on how this form is to be submitted, number of copies, etc., see the instructions applicable to the FHA Application for Mortgage Insurance, VA Request for Determination of Reasonable Value, or other, as the case may be.

2. Describe all materials and equipment to be used, whether or not shown on the drawings, by marking an X in each appropriate check-box and entering the information called for in each space. If space is inadequate, enter "See misc." and describe under item 27 or on an attached sheet. THE USE OF PAINT CONTAINING MORE THAN THE PERCENT OF LEAD BY WEIGHT PERMITTED BY LAW IS PROHIBITED.

3. Work not specifically described or shown will not be considered unless required, then the minimum acceptable will be assumed. Work exceeding minimum requirements cannot be considered unless specifically described.

4. Include no alternates, or equal phrases, or contradictory items. (Consideration of a request for acceptance of substitute materials or equipment is not thereby precluded.)

5. Include signatures required at the end of this form.

6. The construction shall be completed in compliance with the related drawings and specifications, as amended during processing. The specifications include this Description of Materials and the applicable building code.

1. EXCAVATION:
Bearing soil, type Firm clay; Note: Where fill is in excess of 18", concrete piers to be installed at 8' O.C. and the cost will be added to the contract.

2. FOUNDATIONS:
Footings: concrete mix transite 14"x18" ftg ; strength psi 2500 PSI Reinforcing (4) 5/8" steel rebar
Foundation wall: material 2500 PSI concrete concrete Reinforcing ________
Interior foundation wall: material 2500 PSI concrete Party foundation wall 2500 concrete
Columns: material and sizes ________ Piers: material and reinforcing ________
Girders: material and sizes ________ Sills: material W.Coast Utility Douglas Fir w/sill sealer
Basement entrance areaway ________ Window areaways ________
Waterproofing waterproof mix in concrete Footing drains open mortar joints
Termite protection Pretreat soil and stem wall with Chlordane and issue 5 year warranty.
Basementless space: ground cover ________ ; insulation ________ ; foundation vents ________
Special foundations: Thicken slab beam with (2) 5/8" steel rebars to undisturbed soil.
Additional information: See patio details, also porch and fireplace footing.

3. CHIMNEYS:
Material brick Prefabricated *(make and size)* ________
Flue lining: material terra cotta Heater flue size 6" metalbestos Fireplace flue size 12"x12"
Vents *(material and size)*: gas or oil heater ________ ; water heater metalbestos
Additional information: ________

4. FIREPLACES:
Type: ☐ solid fuel; ☒ gas-burning; ☐ circulator *(make and size)* 42"vestal damper Ash dump and clean-out metal
Fireplace: Facing brick ; lining firebrick ; hearth brick ; mantel wood
Additional information: Note: Fireplace stubbed and keyed for gas outlet.

5. EXTERIOR WALLS: 2x4 studs @ 16" O.C. 1/2" plywood at corners
Wood frame: wood grade, and species w.c. utility d.f. ☒ Corner bracing. Building paper or felt 15# felt
Sheathing flintcoat ; thickness 3/4" ; width 24" ; ☒ solid; ☐ space ____ "o.c.; ☐ diagonal; ____
Siding polyurethane ; grade ____ ; type ____ ; size ____ ; exposure ____ " ; fastening ____
Shingles ____ ; grade ____ ; type ____ ; size ____ ; exposure ____ " ; fastening ____
Stucco ____ ; thickness ____ "; Lath ____ ; weight ____ lb.
Masonry veneer face brick Sills brick Lintels steel supporting brick Base flashing ____
Masonry: ☐ solid ☐ faced ☐ stuccoed; total wall thickness ____ " ; facing thickness ____ " ; facing material ____
Backup material ____ ; thickness ____ "; bonding ____
Door sills ____ Window sills ____ Lintels ____ Base flashing ____
Interior surfaces: dampproofing, ____ coats of ____ ; furring ____
Additional information: Two coats of exterior paint or stain as selected by owner.
Exterior painting: material ____ ; number of coats ____
Gable wall construction: ☐ same as main walls; ☒ other construction Cedar plywood or vertical siding as shown on elevations.

6. FLOOR FRAMING:
Joists: wood, grade, and species ____ ; other ____ ; bridging ____ ; anchors 1/2"x8"
Concrete slab: ☐ basement floor; ☐ first floor; ☒ ground supported; ☐ self-supporting; mix 3500 PSI concrete; thickness 4 ",
reinforcing 6x6-W1.4xW1.4 WWF ; insulation 1" perimeter ; membrane waterproof concrete
Fill under slab; material sand fill ; thickness 4 ". Additional information: ____

7. SUBFLOORING: *(Describe underflooring for special floors under item 21.)*
Material: grade and species 3/4" AC exterior plywood ; size ____ ; type ____
Laid: ☐ first floor; ☐ second floor ☒ attic 200 sq. ft; ☐ diagonal; ☐ right angles. Additional information: ____

8. FINISH FLOORING: *(Wood only. Describe other finish flooring under item 21.)*

LOCATION	ROOMS	GRADE	SPECIES	THICKNESS	WIDTH	BLDG. PAPER	FINISH
First floor	living area, formal dining,						
Second floor	hall and bedrooms: pad	and carpet	installed.				allowance per square foot
Attic floor ____ sq. ft.							

Additional information: ____

According to the Paperwork Reduction Act of 1995, an agency may not conduct or sponsor, and a person is not required to respond to a collection of information unless it displays a valid OMB control number. The valid OMB control number for this information collection is 0575-0042. The time required to complete this information collection is estimated to average 15 minutes per response, including the time for reviewing instructions, searching existing data sources, gathering and maintaining the data needed, and completing and reviewing the collection of information.

HUD-FHA 2005
VA Form 26-1852

RD 1924-2 (Rev. 7-99)

104F49A.EPS

Figure 49 ◆ Materials specification. (1 of 4)

9. PARTITION FRAMING:
Studs: wood, grade, and species w.c. utility d.f. size and spacing 2x4 studs @ 16" oc Other _____
Additional information: NOTE: 2x6 studs where needed for plumbing vents.

10. CEILING FRAMING:
2x4 flat and 2x6 up
Joists: wood, grade, and species w.c. utility d.f. Other see floor plan Bridging midspan stiffener
Additional information: See typical sheet for construction details.

11. ROOF FRAMING:
Rafters: wood, grade, and species w.c. utility d.f. Roof trusses (see detail): grade and species _____
Additional information: 2x6 rafters @ 24" oc, unless spans exceed allowable.

12. ROOFING:
Sheathing: wood, grade, and species west coast utility Douglas fir ; ☐ solid ☒ spaced ____ " o.c.
Roofing 240 #3 tab sq. butt ; grade #1 ; size_____ ; type _____
Underlay asphalt shingles ; weight or thickness 5/2 ; size 16" ; fastening nail
Built-up roofing _____ ; number of plies _____ ; surface material _____
Flashing: material galvanized iron ; gage or weight 26 gauge ; ☐ gravel stops; ☐ snow guards
Additional information: _____

13. GUTTERS AND DOWNSPOUTS:
Gutters: material galvanized iron ; gage or weight 26 gauge; size 5" ; shape o-gee
Downspouts: material galvanized iron; gage or weight 26 gauge; size 3"x4" ; shape square ; number as need
Downspouts connected to: ☐ Storm sewer; ☐ sanitary sewer; ☐ dry-well. ☒ Splash blocks: material and size _____
Additional information: _____

14. LATH AND PLASTER:
Lath ☐ walls, ☐ ceilings: material _____ ; weight or thickness_____ Plaster: coats ____ ; finish _____
Dry-wall ☒ walls, ☒ ceilings: material Sheetrock ; thickness 1/2" ; finish medium texture
Joint treatment mud and tape; NOTE: Sheetrock applied according to manufacturer's specs.

15. DECORATING: *(Paint, wallpaper, etc.)*

ROOMS	WALL FINISH MATERIAL AND APPLICATION	CEILING FINISH MATERIAL AND APPLICATION
Kitchen		
Bath	ALL ROOMS: paint and/or wallpaper	all ceilings to be textured
Other	WALLPAPER ALLOWANCE: $	and painted two coats flat paint.

Additional information: PANELING: 1/4" v-groove ash (includes cost of labor)

16. INTERIOR DOORS AND TRIM:
Doors: type hollow core slab ; material ash or birch ; thickness 1-3/8"
Door trim: type detail ; material white pine Base: type detail ; material white pine ; size 3-1/4"
Finish: doors hand-rubbed stain ; trim hand-rubbed stain
Other trim *(item, type and location)* NOTE: All headers to meet schedule as shown on detail sheet.
Additional information: _____

17. WINDOWS: All Thermopane Glass - fixed and operating
Windows: type single hung ; make Alenco or equal ; material aluminum ; sash thickness 1"
Glass: grade DS6 ; ☐ sash weights; ☒ balances, type spring ; head flashing galvanized
Trim: type Sheetrock return ; material Sheetrock Paint texture & paint ; number coats 2
Weatherstripping; type with units ; material wool pile Storm sash, number _____
Screens: ☒ full; ☐ half; type metal or aluminum ; number all ; screen cloth material 16" aluminum mesh
Basement windows: type _____ ; material _____ ; screens, number _____ ; Storm sash, number _____
Special windows See plans for fixed windows
Additional information: NOTE: Sheetrock return on all windows with wood stool and apron.

18. ENTRANCES AND EXTERIOR DETAIL:
Main entrance door: material ash panel ; width see plans ; thickness 1-3/4"; Frame: material w. pine ; thickness 1-3/8"
Other entrance doors: material w.p. m.c. ; width see plans ; thickness 1-3/4"; Frame: material w. pine ; thickness 1-3/8"
Head flashing galvanized iron Weatherstripping: type spring bronze ; Thresholds aluminum
Screen doors: thickness ____" ; number _____ ; screen cloth material _____ Storm doors: thickness ____" ; number _____
Combination storm and screen doors: thickness ____" ; number ____ ; screen cloth material _____
Shutters: ☐ hinged; ☐ fixed. Railings _____ ; Attic louvers galvanized back screen in soffit
Exterior millwork: grade and species redwood or rough cedar Paint exterior paint or stain ; number coats 2
Additional information: Thermopane glass sliding doors with aluminum frame

19. CABINETS AND INTERIOR DETAIL:
Kitchen cabinets, wall units: material ash with white pine stiles ; lineal feet of shelves ____ ; shelf width 12"
Base units: material ash with white pine ; counter top Formica ; edging self-edge Formica
Back and end splash full Formica Finish of cabinets enamel or hand-rubbed stain ; number coats 5
Medicine cabinets: make _____ ; model _____
Other cabinets and built-in furniture Ash raised panel door and drawer fronts -- shop built
Additional information: Built-in vanities with synthetic marble or Formica top and splash; built-in ash bookcases (See Plans)

20. STAIRS:

STAIR	TREADS		RISERS		STRINGS		HANDRAIL		BALUSTERS	
	Material	Thickness	Material	Thickness	Material	Thickness	Material	Thickness	Material	Thickness
Basement										
Main										
Attic										

Disappearing: make and model number pull-down ladder to attic access in garage. 2040
Additional information: _____

HUD-FHA 2005
VA Form 26-1852

2

104F49B.EPS

Figure 49 ◆ Materials specification. (2 of 4)

21. SPECIAL FLOORS AND WAINSCOT: (Describe carpet as listed in Certified Products Directory.)

	Location	Material, Color, Border, Sizes, Gage, Etc.	Threshold Material	Wall Base Material	Underfloor Material
Floors	Kitchen	utility: sheet vinyl			
	Bath	ceramic tile	ceramic	white pine	concrete
	Main entry	earthstone	aluminum	white pine	concrete
	Rear entry	sheet vinyl	aluminum		concrete

	Location	Material, Color, Border, Sizes, Gage, Etc. (tub & shower)	Height shower	Height Over Tub	Height in Showers (From Floor)
Wainscot	Bath	ceramic tile over metl lath and grout area only)		72"	72"
	remainder of				
	bath walls: 1/4" v-groove ash wainscoting 32" high				

Bathroom accessories: ☐ Recessed; material _chrome_ ; number _____ ; ☐ Attached; material _chrome_ ; number _____
(brushed)   (brushed)
Additional information: _____

22. PLUMBING:

Fixture	Number	Location	Make	Mfr's Fixture Identification No.	Size	Color
Sink	1	kitchen	Kohler	Delafield k-5950	32"x21"	color
Lavatory		baths	Kohler	Caxton K-2210	19"x15"	color
Water closet		baths	Kohler	Wellsworth K-3500 eb	reverse trap	color
Bathtub		bath	Kohler	Villager K-715	60"	color
Shower over tubΔ		bath	Kohler	Chrome divertor over tub		
Stall showerΔ		bath	ceramic stall shower with glass door (job built)			
Laundry trays						
Optional:	1	Kohler party bar sink with single lever faucet in utility room				
NOTE: Kohler faucets on all lavatories and sinks.						

Δ ☒ Curtain rod Δ ☒ Door ☐ Shower pan: material _____
Water supply: ☒ public; ☐ community system: ☐ individual (private) system.*
Sewage disposal ☒ public; ☐ community system: ☐ individual (private) system.*
* Show and describe individual system in complete detail in separate drawings and specifications according to requirements.
House drain (inside): ☐ cast iron; ☐ tile; ☒ other _ABS or PVC_ House sewer (outside): ☐ cast iron; ☒ tile; ☐ other _4" ABS_
Water piping: ☐ galvanized steel; ☒ copper tubing; ☐ other _typeI, hard-drawn joints_ Sill cocks, number _frostproof_
Domestic water heater: type _automatic_ ; make and model _Rheem or State_ ; heating capacity _40 gallons_
_25.2_ gph. 100¡ rise. Storage tank: material _glass lined_ ; capacity _40_ gallons.
Gas service: ☒ utility company; ☐ liq. pet. gas; ☐ other _____ Gas piping: ☐ cooking; ☒ house heating.
Footing drains connected to : ☐ storm sewer; ☐ sanitary sewer; ☐ dry well. Sump pump; make and model _____
_____ ; capacity _____ ; discharges into _____

23. HEATING:
☐ Hot water. ☐ Steam. ☐ Vapor. ☐ One-pipe system. ☐ Two-pipe system.
   ☐ Radiators. ☐ Convectors. ☐ Baseboard radiation. Make and model_____
   Radiant panel: ☐ floor; ☐ wall; ☐ ceiling. Panel coil: material_____
   ☐ Circulator. ☐ Return pump. Make and model _____ ; capacity _____ gpm.
   Boiler: make and model _____ Output _____ Btuh.; net rating _____ Btuh.
Additional information: _____
Warm air: ☐ Gravity. ☒ Forced. Type of system _perimeter under floor_
   Duct material: supply _concrete_ ; return _sheet metal_ Insulation _fiberglass_ ; thickness _1"_ ☐ Outside air intake.
   Furnance: make and model _See heating and air conditioning_ Input _120,000_ Btuh.; output _96,000_ Btuh.
   Additional information: _Layout: to be prepared by mechanical contractor._
☐ Space heater; ☐ floor furnace; ☐ wall heater. Input _____ Btuh.; output _____ Btuh.; number units _____
   Make, model _____ Additional information: _____
Controls: make and types _____
Additional information: _____
Fuel: ☐ Coal; ☐ oil; ☐ gas; ☐ liq. pet. gas; ☐ electric; ☐ other _____ ; storage capacity _____
   Additional information: _____
Firing equipment furnished separately: ☐ Gas burner, conversion type. ☐ Stoker: hopper feed ☐ ; bin feed ☐
   Oil burner: ☐ pressure atomizing; ☐ vaporizing _____
   Make and model _____ Control _____ _furnace._
   Additional information: _NOTE: Gas outlet for hot water tank, furnace and fireplace jet, and garage_
Electric heating system: type _____ Input ____ watts;@ _____ volts; output _____ Btuh.
   Additional information: _____
Ventilating equipment: attic fan, make and model _____ ; capacity _7,800_ cfm.
   kitchen exhaust fan, make and model _Kitchen vent hood and exhaust fan._
Other heating, ventilating, or cooling equipment _Nutone or equal ceiling heaters in baths._

24. ELECTRIC WIRING: allow for 200 amp service   as required
Service: ☐ overhead; ☐ underground. Panel: ☐ fuse box; ☒ circuit-breaker; make _square D_ AMPs _____ No. circuits _____
Wiring: ☐ conduit; ☒ armored cable; ☐ nonmetallic cable; ☐ knob and tube; ☐ other _romex wiring if code allows_
Special outlets: ☒ range; ☐ water heater; ☐ other _220 volt for oven, range, dryer, and air conditioning_
☐ Doorbell. ☒ Chimes. Push-button locations _____ Additional information: _____

25. LIGHTING FIXTURES:
Total number of fixtures _see plans_ Total allowance for fixtures, typical installation, $ _____
Nontypical installation _____
Additional information: _Optional Fluorescent grid, dropped ceiling in kitchen as shown on plans_

HUD-FHA 2005   3   **DESCRIPTION OF MATERIALS**
VA Form 26-1852

104F49C.EPS

Figure 49 ◆ Materials specification. (3 of 4)

26. INSULATION:

Location	Thickness	Material, Type, and Method of Installation	Vapor Barrier
Roof			
Ceiling	12"	loose-fill fiberglass insulation in ceiling of house only	(blown between jsts)
Wall	4"	batts- fiberglass insulation- exterior walls of house	(stapled to studs)
Floor	1"	perimeter insulation under slab.	

27. MISCELLANEOUS: *(Describe any main dwelling materials, equipment, or construction items not shown elsewhere; or use to provide additional information where the space provided was inadequate. Always reference by item number to correspond to numbering used on this form.)*

   LIGHTING FIXTURES . . . . $ as per contract

   WALLPAPER . . . . $ per contract labor and material

   LANDSCAPING . . . . $ as per contract

   CARPET ALLOWANCE . . . . $ PER SQUARE YARD INSTALLED as per contract

   Cabinet over washer and dryer.

   Waterproof outlet on patio.

   Ornamental fence and gate at main entrance

HARDWARE: *(make, material, and finish.)*

   Dexter or equal passage sets and locks- all exterior doors keyed alike

   deadbolt locks on all exterior doors.

SPECIAL EQUIPMENT: *(State material or make, model and quantity. Include only equipment and appliances which are acceptable by local law, custom and applicable FHA standards. Do not include items which, by established custom, are supplied by occupant and removed when he vacates premises or chattles prohibited by law from becoming realty.)*

   APPLIANCE ALLOWANCE: $

        AS SELECTED BY OWNER WITHIN SPECIFIED ALLOWANCE

PORCHES:

   foundation construction: 6"x8" poured monolithic with the slab.

   floor: 4" concrete slab with 6x6-W1.4xW1.4 WWF smooth trowel finish.

TERRACES:

   stoops: 4" concrete slab- smooth trowel finish.

   patio: 4" concrete slab- smooth trowel finish. - see plans for size.

GARAGES: automatic garage door opener

   foundation: 14"x 18" concrete footing with (4) 5/8" rebars; 6" concrete stem wall

   floor: 4" concrete with 6x6-W1.4xW1.4 WWF; smooth trowel finish floors.

   interior: 3/8" prefinished Sheetrock on walls; texture and paint 1/2" Sheetrock on ceiling

WALKS AND DRIVEWAYS:

Driveway: width plot (see) ; base material tamped earth ; thickness 4 "; surfacing material concrete ; thickness 4 "

Front walk: width 36" ; material concrete ; thickness 4 ". Service walk: width_____ ; material _____ ; thickness _____ "

Steps: material _____ ; treads _____ "; risers _____ ". Check walls _____

OTHER ONSITE IMPROVEMENTS:

(Specify all exterior onsite improvements not described elsewhere, including items such as unusual grading, drainage structures, retaining walls, fence railings, and accessory structures.)

   NOTE: All dimensions to be rechecked on site prior to beginning construction by

   builder and builder shall be responsible for the same.

LANDSCAPING, PLANTING, AND FINISH GRADING:

Topsoil _____ " thick: ☐ front yard; ☐ side yards; ☐ rear yard to _____ feet behind main building.

Lawns *(seeded, sodded, or sprigged):*. ☐ front yard _____ ; ☐ side yards _____ ; ☐ rear yard _____

Planting: ☐ as specified and shown on drawings; ☐ as follows:

_____ Shade trees, decidous. _____ " caliper.   _____ Evergreen trees _____ ' to _____ ', B & B.

_____ Low flowering trees, decidous, _____ ' to _____ '   _____ Evergreen shrubs _____ ' to _____ ', B & B.

_____ High-growing shrubs, decidous, _____ ' to _____ '   _____ Vines, 2-year _____

_____ Medium-growing shrubs, decidous, _____ ' to _____ '   _____

_____ Low-growing shrubs, decidous, _____ ' to _____ '

IDENTIFICATION —'This exhibit shall be identified by the signature of the builder, or sponsor, and/or the proposed mortgagor if the latter is known at the time of application.

Date _____   Signature _____

        Signature _____

HUD-FHA 2005
VA Form 26-1852   4

104F49D.EPS

Figure 49 ◆ Materials specification. (4 of 4)

6.1.2 Technical Aspects

The technical aspects section includes information on materials that are specified by standard numbers and by standard national testing organizations such as the American Society for Testing and Materials International (ASTM). The technical data section of specifications can be any of three types:

- *Outline specifications* – These specifications list the materials to be used in order of the basic parts of the job, such as foundation, floors, and walls.

- *Fill-in specifications* – This is a standard form filled in with pertinent information. It is typically used on smaller jobs.

- *Complete specifications* – For ease of use, most specifications written for large construction jobs are organized in the Construction Specification Institute format called the Uniform Construction Index. This is known as the CSI format and is explained in the next section.

6.2.0 CSI Format

The Construction Specification Institute (CSI) developed the Uniform Construction Index. It allows all specifications, product information, and cost data to be arranged into a uniform system. This format is now used for most large construction projects in North America. All construction is divided into sixteen divisions, and each division has several sections and subsections (*Figure 50*). The following outline briefly describes the various divisions normally included in a set of specifications written to the CSI format:

- *Division 1 – General Requirements:* Covers the general and special conditions by which the whole job is governed. It summarizes the work, alternatives, project meetings, submissions, quality control, temporary facilities and controls, products, and project closeout. Every responsible person involved with the project should become familiar with this division.

- *Division 2 – Site Work:* Outlines work involving such items as paving, sidewalks, outside utility lines (electrical, plumbing, gas, and telephone), landscaping, grading, and other items pertaining to the outside of the building.

- *Division 3 – Concrete:* Covers work involving footings, concrete formwork, expansion and contraction joints, cast-in-place concrete, specially finished concrete, precast concrete, and concrete slabs.

- *Division 4 – Masonry:* Covers concrete, mortar, stone, and masonry accessories.

- *Division 5 – Metals:* Covers metal roofs, structural metal framing, metal joists, metal decking, ornamental metal, and expansion control.

- *Division 6 – Wood and Plastics:* Includes rough carpentry, heavy timber construction, trestles, prefabricated structural wood, finish carpentry, wood treatment, architectural woodwork, and plastic fabrications.

- *Division 7 – Thermal and Moisture Protection:* Waterproofing is the main topic discussed under this division. Other related items such as building insulation, shingles, roofing tiles, preformed roofing and siding, membrane roofing, sheet metal work, wall flashing, roof accessories, and sealants are also included.

- *Division 8 – Doors and Windows:* Includes all types of doors and frames such as metal, plastic, or wood. Windows and framing are also included, along with hardware and other window and door accessories.

- *Division 9 – Finishes:* Includes the types, quality, and workmanship of lath and plaster, gypsum wallboard, tile, terrazzo, acoustical treatment, ceiling suspension systems, wood flooring, floor treatment, special coatings, painting, and wallcoverings.

- *Division 10 – Specialties:* Covers specialty items such as chalkboards and tackboards, compartments and cubicles, louvers and vents that are not connected with the HVAC system, wall and corner guards, access flooring, specialty modules, pest control, fireplaces, flagpoles, identifying devices, lockers, protective covers, postal specialties, partitions, scales, storage shelving, and wardrobe specialties.

- *Division 11 – Equipment:* Includes central vacuuming systems, bank vaults, darkrooms, food service, vending machines, laundry equipment, and many similar items.

- *Division 12 – Furnishings:* Includes items such as cabinets and storage, fabrics, furniture, rugs and mats, seating, and similar furnishings.

- *Division 13 – Special Construction:* Includes air-supported structures, incinerators, and other special items.

- *Division 14 – Conveying Systems:* Covers conveying apparatus such as dumbwaiters, elevators, hoists, cranes, lifts, material-handling systems, turntables, moving stairs and walks, pneumatic tube systems, and powered scaffolding.

- *Division 15 – Mechanical:* Includes plumbing, HVAC, and related work. Electric heat is sometimes covered under Division 16, especially if individual baseboard heating units are used in each room or area of the building.

- *Division 16 – Electrical:* Covers all electrical requirements for the building, including lighting, power, alarm, and communications systems. It also covers special electrical systems and related electrical equipment.

DIVISIONS

DIVISION 1 – GENERAL REQUIREMENTS

    2 – SITE WORK

    3 – CONCRETE

    4 – MASONRY

    5 – METALS

    6 – WOOD AND PLASTICS

    7 – THERMAL AND MOISTURE PROTECTION

    8 – DOORS AND WINDOWS

    9 – FINISHES

    10 – SPECIALTIES

    11 – EQUIPMENT

    12 – FURNISHINGS

    13 – SPECIAL CONSTRUCTION

    14 – CONVEYING SYSTEMS

    15 – MECHANICAL

    16 – ELECTRICAL

DIVISION 1 – GENERAL REQUIREMENTS

01010 SUMMARY OF WORK
01020 ALLOWANCES
01025 MEASUREMENT AND PAYMENT
01030 ALTERNATES/ALTERNATIVES
01035 MODIFICATION PROCEDURES
01040 COORDINATION
01050 FIELD ENGINEERING
01060 REGULATORY REQUIREMENTS
01070 IDENTIFICATION SYSTEMS
01090 REFERENCES
01100 SPECIAL PROJECT PROCEDURES
01200 PROJECT MEETINGS
01300 SUBMITTALS
01400 QUALITY CONTROL
01500 CONSTRUCTION FACILITIES AND TEMPORARY
      CONTROLS
01600 MATERIAL AND EQUIPMENT
01650 FACILITY STARTUP/COMMISSIONING
01700 CONTRACT CLOSEOUT
01800 MAINTENANCE

DIVISION 2 – SITE WORK

02010 SUBSURFACE INVESTIGATION
02050 DEMOLITION
02100 SITE PREPARATION
02140 DEWATERING
02150 SHORING AND UNDERPINNING
02160 EXCAVATION SUPPORT SYSTEMS
02170 COFFERDAMS
02200 EARTHWORK
02300 TUNNELING
02350 PILES AND CAISSONS
02450 RAILROAD WORK
02480 MARINE WORK
02500 PAVING AND SURFACING
02600 UTILITY PIPING MATERIALS
02660 WATER DISTRIBUTION
02680 FUEL AND STEAM DISTRIBUTION
02700 SEWERAGE AND DRAINAGE
02760 RESTORATION OF UNDERGROUND PIPE
02770 PONDS AND RESERVOIRS
02780 POWER AND COMMUNICATIONS
02800 SITE IMPROVEMENTS
02900 LANDSCAPING

DIVISION 3 – CONCRETE

03100 CONCRETE FORMWORK
03200 CONCRETE REINFORCEMENT
03250 CONCRETE ACCESSORIES
03300 CAST-IN-PLACE CONCRETE
03370 CONCRETE CURING
03400 PRECAST CONCRETE
03500 CEMENTITIOUS DECKS AND TOPPINGS
03600 GROUT
03700 CONCRETE RESTORATION AND CLEANING
03800 MASS CONCRETE

DIVISION 4 – MASONRY

04100 MORTAR AND MASONRY GROUT
04150 MASONRY ACCESSORIES
04200 UNIT MASONRY
04400 STONE
04500 MASONRY RESTORATION AND CLEANING
04550 REFRACTORIES
04600 CORROSION RESISTANT MASONRY
04700 SIMULATED MASONRY

DIVISION 5 – METALS

05010 METAL MATERIALS
05030 METAL COATINGS
05050 METAL FASTENING
05100 STRUCTURAL METAL FRAMING
05200 METAL JOISTS
05300 METAL DECKING
05400 COLD-FORMED METAL FRAMING
05500 METAL FABRICATIONS
05580 SHEET METAL FABRICATIONS
05700 ORNAMENTAL METAL
05800 EXPANSION CONTROL

DIVISION 6 – WOOD AND PLASTICS

06050 FASTENERS AND ADHESIVES
06100 ROUGH CARPENTRY
06130 HEAVY TIMBER CONSTRUCTION
06150 WOOD AND METAL SYSTEMS
06170 PREFABRICATED STRUCTURAL WOOD
06200 FINISH CARPENTRY
06300 WOOD TREATMENT
06400 ARCHITECTURAL WOODWORK
06500 STRUCTURAL PLASTICS
06600 PLASTIC FABRICATIONS
06650 SOLID POLYMER FABRICATIONS

DIVISION 7 – THERMAL AND MOISTURE PROTECTION

07100 WATERPROOFING
07150 DAMPPROOFING
07180 WATER REPELLENTS
07190 VAPOR RETARDERS
07195 AIR BARRIERS
07200 INSULATION
07240 EXTERIOR INSULATION AND FINISH SYSTEMS
07250 FIREPROOFING
07270 FIRESTOPPING
07300 SHINGLES AND ROOFING TILES
07400 MANUFACTURED ROOFING AND SIDING
07480 EXTERIOR WALL ASSEMBLIES
07500 MEMBRANE ROOFING
07570 TRAFFIC COATINGS
07600 FLASHING AND SHEET METAL
07700 ROOF SPECIALTIES AND ACCESSORIES
07800 SKYLIGHTS
07900 JOINT SEALERS

DIVISION 8 – DOORS AND WINDOWS

08100 METAL DOORS AND FRAMES
08200 WOOD AND PLASTIC DOORS
08250 DOOR OPENING ASSEMBLIES
08300 SPECIAL DOORS
08400 ENTRANCES AND STOREFRONTS
08500 METAL WINDOWS
08600 WOOD AND PLASTIC WINDOWS
08650 SPECIAL WINDOWS
08700 HARDWARE
08800 GLAZING
08900 GLAZED CURTAIN WALLS

DIVISION 9 – FINISHES

09100 METAL SUPPORT SYSTEMS
09200 LATH AND PLASTER
09250 GYPSUM BOARD
09300 TILE
09400 TERRAZZO
09450 STONE FACING
09500 ACOUSTICAL TREATMENT
09540 SPECIAL WALL SURFACES
09545 SPECIAL CEILING SURFACES
09550 WOOD FLOORING
09600 STONE FLOORING
09630 UNIT MASONRY FLOORING
09650 RESILIENT FLOORING
09680 CARPET
09700 SPECIAL FLOORING
09780 FLOOR TREATMENT
09800 SPECIAL COATINGS
09900 PAINTING
09950 WALL COVERINGS

104F50A.EPS

Figure 50 ◆ Format of CSI specifications. (1 of 2)

DIVISIONS

DIVISION 1 – GENERAL REQUIREMENTS

2 – SITE WORK

3 – CONCRETE

4 – MASONRY

5 – METALS

6 – WOOD AND PLASTICS

7 – THERMAL AND MOISTURE PROTECTION

8 – DOORS AND WINDOWS

9 – FINISHES

10 – SPECIALTIES

11 – EQUIPMENT

12 – FURNISHINGS

13 – SPECIAL CONSTRUCTION

14 – CONVEYING SYSTEMS

15 – MECHANICAL

16 – ELECTRICAL

DIVISION 10 – SPECIALTIES

10100 VISUAL DISPLAY BOARDS
10150 COMPARTMENTS AND CUBICLES
10200 LOUVERS AND VENTS
10240 GRILLES AND SCREENS
10250 SERVICE WALL SYSTEMS
10260 WALL AND CORNER GUARDS
10270 ACCESS FLOORING
10290 PEST CONTROL
10300 FIREPLACES AND STOVES
10340 MANUFACTURED EXTERIOR SPECIALTIES
10350 FLAGPOLES
10400 IDENTIFYING DEVICES
10450 PEDESTRIAN CONTROL DEVICES
10500 LOCKERS
10520 FIRE PROTECTION SPECIALTIES
10530 PROTECTIVE COVERS
10550 POSTAL SPECIALTIES
10600 PARTITIONS
10650 OPERABLE PARTITIONS
10670 STORAGE SHELVING
10700 EXTERIOR PROTECTION DEVICES FOR
      OPENINGS
10750 TELEPHONE SPECIALTIES
10800 TOILET AND BATH ACCESSORIES
10880 SCALES
10900 WARDROBE AND CLOSET SPECIALTIES

DIVISION 11 – EQUIPMENT

11010 MAINTENANCE EQUIPMENT
11020 SECURITY AND VAULT EQUIPMENT
11030 TELLER AND SERVICE EQUIPMENT
11040 ECCLESIASTICAL EQUIPMENT
11050 LIBRARY EQUIPMENT
11060 THEATER AND STAGE EQUIPMENT
11070 INSTRUMENTAL EQUIPMENT
11080 REGISTRATION EQUIPMENT
11090 CHECKROOM EQUIPMENT
11100 MERCANTILE EQUIPMENT
11110 COMMERCIAL LAUNDRY AND DRY CLEANING
      EQUIPMENT
11120 VENDING EQUIPMENT
11130 AUDIOVISUAL EQUIPMENT
11140 VEHICLE SERVICE EQUIPMENT
11150 PARKING CONTROL EQUIPMENT
11160 LOADING DOCK EQUIPMENT
11170 SOLID WASTE HANDLING EQUIPMENT
11190 DETENTION EQUIPMENT
11200 WATER SUPPLY AND TREATMENT EQUIPMENT
11280 HYDRAULIC GATES AND VALVES
11300 FLUID WASTE TREATMENT AND DISPOSAL
      EQUIPMENT
11400 FOOD SERVICE EQUIPMENT
11450 RESIDENTIAL EQUIPMENT
11460 UNIT KITCHENS
11470 DARKROOM EQUIPMENT
11480 ATHLETIC, RECREATIONAL, AND THERAPEUTIC
      EQUIPMENT

DIVISION 11 – EQUIPMENT CONT.

11500 INDUSTRIAL AND PROCESS EQUIPMENT
11600 LABORATORY EQUIPMENT
11650 PLANETARIUM EQUIPMENT
11660 OBSERVATORY EQUIPMENT
11680 OFFICE EQUIPMENT
11700 MEDICAL EQUIPMENT
11780 MORTUARY EQUIPMENT
11850 NAVIGATION EQUIPMENT
11870 AGRICULTURAL EQUIPMENT

DIVISION 12 – FURNISHINGS

12050 FABRICS
12100 ARTWORK
12300 MANUFACTURED CASEWORK
12500 WINDOW TREATMENTS
12600 FURNITURE AND ACCESSORIES
12670 RUGS AND MATS
12700 MULTIPLE SEATING
12800 INTERIOR PLANTS AND PLANTERS

DIVISION 13 – SPECIAL CONSTRUCTION

13010 AIR-SUPPORTED STRUCTURES
13020 INTEGRATED ASSEMBLIES
13030 SPECIAL-PURPOSE ROOMS
13080 SOUND, VIBRATION, AND SEISMIC CONTROL
13090 RADIATION PROTECTION
13100 NUCLEAR REACTORS
13120 PRE-ENGINEERED STRUCTURES
13150 AQUATIC FACILITIES
13175 ICE RINKS
13180 SITE-CONSTRUCTED INCINERATORS
13185 KENNELS AND ANIMAL SHELTERS
13200 LIQUID AND GAS STORAGE TANKS
13220 FILTER UNDERDRAINS AND MEDIA
13230 DIGESTER COVERS AND APPURTENANCES
13240 OXYGENATION SYSTEMS
13260 SLUDGE CONDITIONING SYSTEMS
13300 UTILITY CONTROL SYSTEMS
13400 INDUSTRIAL AND PROCESS CONTROL
      SYSTEMS
13500 RECORDING INSTRUMENTATION
13550 TRANSPORTATION CONTROL
      INSTRUMENTATION
13600 SOLAR ENERGY SYSTEMS
13700 WIND ENERGY SYSTEMS
13750 COGENERATION SYSTEMS
13800 BUILDING AUTOMATION SYSTEMS
13900 FIRE SUPPRESSION AND SUPERVISORY
      SYSTEMS
13950 SPECIAL SECURITY CONSTRUCTION

DIVISION 14 – CONVEYING SYSTEMS

14100 DUMBWAITERS
14200 ELEVATORS
14300 ESCALATORS AND MOVING WALKS
14400 LIFTS
14500 MATERIAL HANDLING SYSTEMS
14600 HOISTS AND CRANES
14700 TURNTABLES
14800 SCAFFOLDING
14900 TRANSPORTATION SYSTEMS

DIVISION 15 – MECHANICAL

15050 BASIC MECHANICAL MATERIALS AND
      METHODS
15250 MECHANICAL INSULATION
15300 FIRE PROTECTION
15400 PLUMBING
15500 HEATING, VENTILATING, AND AIR
      CONDITIONING
15550 HEAT GENERATION
15650 REFRIGERATION
15750 HEAT TRANSFER
15850 AIR HANDLING
15880 AIR DISTRIBUTION
15950 CONTROLS
15990 TESTING, ADJUSTING, AND BALANCING

DIVISION 16 – ELECTRICAL

16050 BASIC ELECTRICAL MATERIALS AND METHODS
16200 POWER GENERATION/BUILT-UP SYSTEMS
16300 MEDIUM VOLTAGE DISTRIBUTION
16400 SERVICE AND DISTRIBUTION
16500 LIGHTING
16600 SPECIAL SYSTEMS
16700 COMMUNICATIONS
16850 ELECTRIC RESISTANCE HEATING
16900 CONTROLS
16950 TESTING

104F50B.EPS

Figure 50 ◆ Format of CSI specifications. (2 of 2)

Currently, CSI-formatted specifications consist of 16 divisions, as described in this module. However, there is some consensus in the construction industry that there is a need to add a Division 17. This division would cover telecommunications and building automation systems and ensure that these systems are taken into consideration during the design phase of a building.

The proposed subdivisions under Division 17 are as follows:

- General requirements
- Cable plant
- Local area network (LAN)
- Audio/video
- Wide area network (WAN)
- Architectural, electrical, and mechanical systems
- Intrabuilding communication systems
- Building automation and control
- Security, access, and surveillance

7.0.0 ◆ BUILDING CODES

Building codes are a vital part of the building industry. They establish the minimum standards required to protect the health, safety, and welfare of the building occupants. Building codes govern the design and specifications for a building in a particular area.

There are several model or suggested building codes used in the United States that have been developed jointly by building officials and industry representatives from all parts of the country. These codes typically serve as the basis for the building codes and laws that govern construction in many states and local communities. Among the most widely used model building codes and the organizations that publish them are the following:

- *Standard Building Code (SBC)* – Published by the Southern Building Code Congress.
- *BOCA National Building Code (NBC)* – Published by the Building Officials and Code Administrators.
- *Uniform Building Code (UBC)* – Published by the International Conference of Building Officials.

In 2000, these three organizations merged into the International Code Council and issued the International Building Code (IBC). This model code has now been adopted throughout most of the United States.

When states, counties, and cities adopt a model code as the basis for their code, they often change it to meet local conditions. They might add further restrictions, or they might only adopt part of the model code. An important general rule to remember about codes is that in almost every case the most stringent local code will apply.

You should be aware of the federal laws, building codes, and restrictions that affect the specific job being constructed. This should also include a basic understanding of the codes that pertain mainly to other trades, such as the National Fire Protection Association (NFPA) gas and electrical codes.

8.0.0 ◆ QUANTITY TAKEOFFS

The **quantity takeoff,** or takeoff procedure, involves surveying, measuring, and counting all materials indicated on a set of drawings. Normally, standard takeoff sheets are available for various trades to use for accurate takeoffs. The sheets are useful because they provide standardization, continuity, and a permanent record. In addition, they reduce the workload as well as the potential for error. In order to make measuring, counting, and calculating tasks easier when performing a takeoff, the following materials will help:

- Colored pencils for checking off items on drawings as they are taken off. (The same color can be used to check off all similar materials.)
- Two drafting scales: one architect's scale and one engineer's scale
- Calculator
- Magnifying glass for examining details on the drawings

Following a standard procedure when doing a takeoff helps ensure accuracy and completeness. Generally, the steps for performing a takeoff are as follows:

Step 1 Determine the scope of work to be done.
- Start with the plans. Leaf through the drawings to determine the following:
  - General location
  - Job-site conditions
  - Size of project
  - Type(s) of work to be done
  - For whom the work is being done (architect, owner, etc.)
  - Whether all the drawings are included in your set
  - Date of the drawings and revision numbers
- Begin to fill out your takeoff sheets using these data.
- Write down a summary of important aspects of the specifications on the takeoff sheets.

Step 2   Begin a quantity takeoff.

- Review plan sheets in sequence:
  - Plot plan
  - Foundation plan
  - Floor plans
  - Exterior elevations
  - Section and detail views
  - Framing plans (if applicable)
  - Door, window, and finish schedules
- Use a systematic approach by working through the structure in a consistent order, such as from foundation to top floor, from front to rear, or from exterior roof line to ground level.
- Review related drawings:
  - Plot and landscaping
  - Mechanical, plumbing, fire sprinklers, and electrical
  - Structural steel
  - Plan sheets
  - Written notes and unusual features

Step 3   Review specifications and seek clarification if necessary.

- Read the specifications again in light of your takeoff. Review the schedules and written notes on the drawings to see if they are in agreement with the written specifications.
- Resolve any questions with your supervisor, the architect, and/or the owner, then record the answers.
- As applicable, discuss the division of responsibilities with your supervisor, architect, and/or the owner, then record these decisions.

Step 4   Determine total quantities for each category of materials.

- Make calculations and review for errors.
- Check off and summarize the total quantity needed for each item of material. Remember to adjust material quantities for normal units of purchase and any unusual waste considerations.

Summary

Construction drawings (blueprints) are drawings that show skilled tradespeople how to build a specific building or structure. Specifications are written instructions provided by architectural and engineering firms to the general contractor and, consequently, to the subcontractors. Specifications are just as important as the drawings in a set of plans. They furnish what the drawings cannot in that they define the quality of work to be done and the materials to be used. When there is a conflict between the drawings and the specifications, the specifications normally take precedence.

Review Questions

1. What is the elevation of the point of beginning for the site plan shown in *Figure 4*?
   a. 539.05'
   b. 540.85'
   c. 551.12'
   d. 552.92'

2. How far apart are the anchor bolts used in the pier footings shown in foundation plan *Figure 5*?
   a. 6" center to center
   b. 8" center to center
   c. 10" center to center
   d. 12" center to center

3. A drawing that shows the horizontal view of a building, the length and width of the building, and the floor layout of the rooms is called a(n) _____.
   a. elevation drawing
   b. foundation plan
   c. plot plan
   d. floor plan

4. The distance between the centers of the rear window and door in the garage shown on the floor plan in *Figure 7* is _____.
   a. 6'-0"
   b. 6'-4"
   c. 8'-0"
   d. 10'-0"

5. How many roof drains are called for in the roof plan shown in *Figure 8*?
   a. Three
   b. Four
   c. Five
   d. Six

6. The vertical mansard (type of roof) shown in the elevation drawing on *Figure 9* is made of _____.

    a. 26 gauge, Type SR arctic white metal
    b. 26 gauge, Type SS arctic white metal
    c. 8" vertical score block
    d. 8" split rib block

7. The elevation of the bottom of the vertical mansard as shown in the section drawing *Figure 10* is _____.

    a. 95.00'
    b. 104.33'
    c. 107.00'
    d. 108.75'

8. Item F listed in the grille, register, and diffuser schedule shown on *Figure 14* identifies a _____ diffuser that delivers _____.

    a. 12" × 12"; 240 cfm
    b. 12" × 12"; 250 cfm
    c. 18" × 18"; 235 cfm
    d. 18" × 18"; 280 cfm

9. As shown on the plumbing plan in *Figure 15*, _____ urinal(s) must be installed.

    a. one
    b. two
    c. four
    d. six

10. What size chilled-water supply piping should be connected to air handling unit 1 shown in the mechanical plan on *Figure 17*?

    a. 1¼"
    b. 1½"
    c. 2"
    d. 2½"

11. The electrical plan in *Figure 18* shows the new power distribution panel, panel D, to be installed in _____.

    a. hallway 125
    b. lobby 123
    c. room 126
    d. room 131

12. Contour lines are used on site/plot plans to show _____ elevations.

    a. locations of roads and railroad tracks
    b. locations of underground utility lines
    c. existing and finished grade
    d. the outline of proposed building

Refer to *Figure 1* when answering Questions 13 through 15.

13. When used on site (plot) plans, the symbol marked H in the figure represents a _____.

    a. property corner
    b. bench mark
    c. required spot elevation
    d. storm sewer

14. When used on drawings, the line marked A in the figure is called a(n) _____ line.

    a. object
    b. telephone line
    c. cutting plane
    d. break

15. The dimension marked I in the figure represents a measurement made from _____.

    a. wall to wall
    b. center to center
    c. outside to center
    d. outside to outside

16. A type of drawing that is made by extending perpendicular (right-angle) lines from an object to create projected plan views is called an isometric drawing.

    a. True
    b. False

17. Anchor bolt placing drawings are considered to be a form of _____ drawing.

    a. civil
    b. structural
    c. mechanical
    d. shop

18. What is the piece mark for the lower column intersected by grid lines E and 3 shown on the erection diagram in *Figure 39*?

    a. A2
    b. B8
    c. C3
    d. C10

19. Drawings that show the types of end connections on beams and columns are _____ drawings.

    a. miscellaneous
    b. sections and details
    c. elevation
    d. shop fabrication

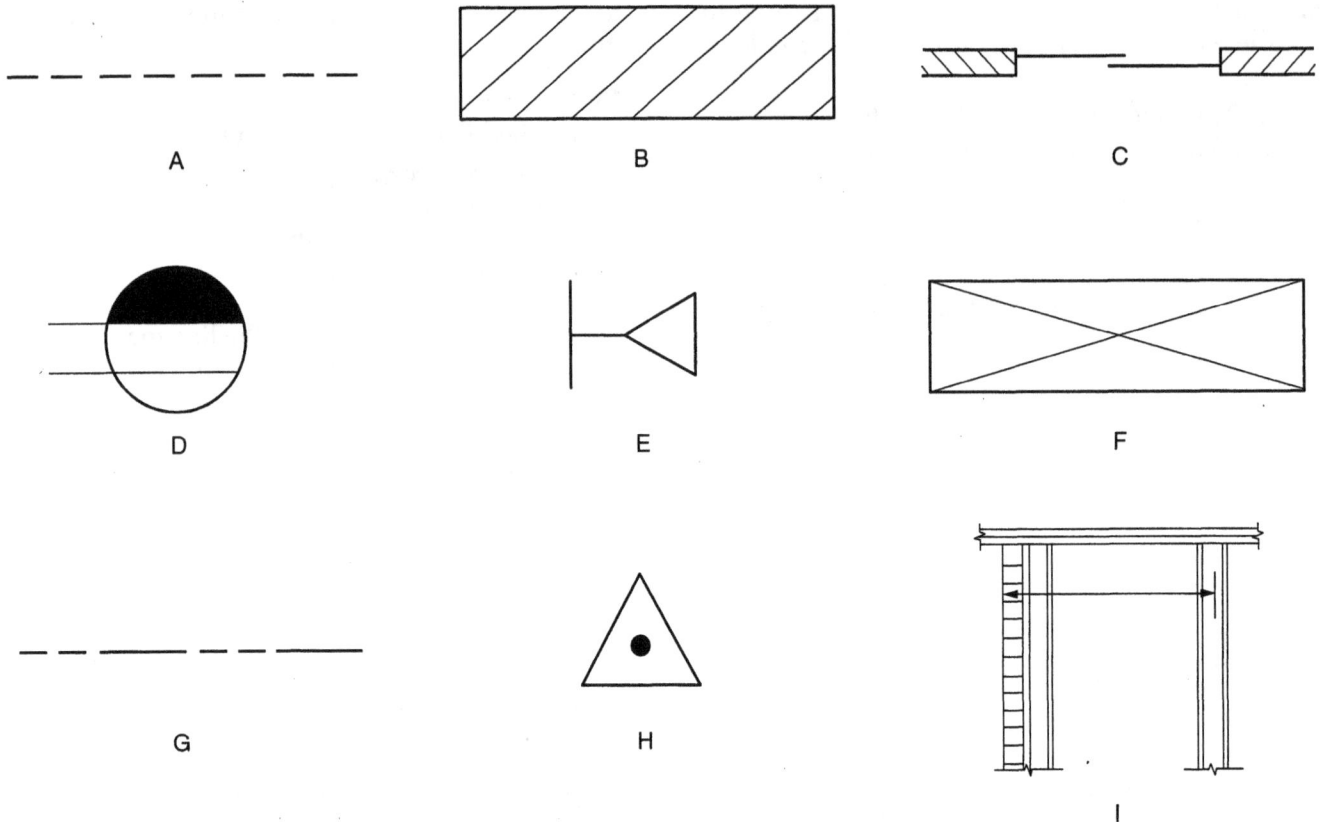

A

B

C

D

E

F

G

H

I

104RQ01.EPS

Figure 1

20. A steel channel has a designation of C12 × 30. This means it is a(n) _____.
    a. American Standard channel with a depth of 12"
    b. miscellaneous channel with a weight of 12 pounds per foot
    c. American Standard channel with a weight of 12 pounds per foot
    d. miscellaneous channel with a weight of 30 pounds per foot

21. The diameter of a #8 rebar is approximately _____.
    a. ½" (0.500 in.)
    b. ¾" (0.750 in.)
    c. 1" (1.000 in.)
    d. 1¼" (1.270 in.)

22. The special and general conditions portion of a specification covers the _____ aspects of contractual arrangements.
    a. non-technical
    b. technical
    c. equipment and material
    d. both non-technical and technical

23. The specification shown in *Figure 49* indicates the size and spacing of partition studs to be _____.
    a. 2 × 4 wood studs at 16" OC
    b. 2 × 4 wood studs at 24" OC
    c. 2 × 4 16-gauge metal studs at 16" OC
    d. 2 × 4 20-gauge metal studs at 24" OC

24. When given specifications written to the CSI format, you would expect to find information on earthwork in Division _____.
    a. 1
    b. 2
    c. 10
    d. 13

25. When doing a quantity takeoff, the first step is to _____.
    a. perform the quantity takeoff
    b. total the materials per category of material
    c. determine the scope of work to be done
    d. review the specifications and seek clarification if necessary

Trade Terms Introduced in This Module

Change order: A formal instruction describing and authorizing a project change.

Contour lines: Imaginary lines on a site/plot plan that connect points of the same elevation. Contour lines never cross each other.

Easement: A legal right-of-way provision on another person's property (for example, the right of a neighbor to build a road or a public utility to install water and gas lines on the property). A property owner cannot build on an area where an easement has been identified.

Elevation view: A drawing giving a view from the front or side of a structure.

Erection drawing: A floor plan-type drawing that identifies and shows the locations of all steel structural members.

Fabricator: A person who provides detailed drawings for the fabrication of components and who fabricates them in a shop for later installation at the job site.

Front setback: The distance from the property line to the front of the building.

Kips: Abbreviation for 1,000 (kilo) pounds.

Isometric drawing: A three-dimensional type drawing in which the object is tilted so that all three faces are equally inclined to the picture plane.

Monuments: Physical structures that mark the locations of survey points.

Nominal size: Approximate or rough size (commercial size) by which lumber, block, etc., is commonly known and sold (for example, 2 × 4). The nominal size is normally slightly larger than the actual size.

Orthographic projection drawing: A multi-view drawing that shows all of the features of an object. The different views are projected at right angles (90°) to each other.

Plan view: A drawing that represents a view looking down on an object.

Property lines: The recorded legal boundaries of a piece of property.

Quantity takeoff: A procedure that involves surveying, measuring, and counting all materials indicated on a set of drawings.

Request for information: A form used to question discrepancies on the drawings or to ask for clarification.

Riser diagram: A type of isometric drawing that depicts the layout, components, and connections of a piping system.

Common Construction Abbreviations

Above mean sea level	ABMSL	Edge of gutter	EG
Abutment	abt.	Edge of pavement	EP
Approximate	approx.	Edge of shoulder	ES
At	@	Elevation	el.
Avenue	Ave.	End wall	EW
Average	avg.	Equation	eqn.
Back of sidewalk	BSW	Existing	exist.
Back of walk	BW	Expressway	Exwy.
Backsight	BS	Fahrenheit	F
Begin curb return	BCR	Fence	fe.
Bench mark	BM	Fence post	FP
Between	betw.	Feet	ft.
Bottom	bot.	Field book	FB
Boulevard	Blvd.	Fill	f
Boundary	bndry.	Finish grade	FG
Bridge	br.	Fire hydrant	FH
Calculated	calc.	Flow line	FL
Cast-iron pipe	CIP	Foot	ft.
Catch basin	CB	Footing	ftg.
Catch point	CP	Foresight	FS
Cement-treated base	CTB	Found	fd.
Concrete block wall	CBW	Foundation	fdn.
Construction	const.	Freeway	Fwy.
Control point	CP	Galvanized	galv.
County	Co.	Galvanized steel pipe	GSP
Court	Ct.	Gas line	GL
Creek	cr.	Gas valve	GV
Curb	cb.	Geodetic	geod.
Curb and gutter	C&G	Grid	grd.
Cut	C	Ground	grnd.
Description	desc.	Gutter	gtr.
Destroyed	dest.	Head wall	hdwl.
Detour	det.	Height	ht.
Direct	D	Height of instrument	HI
Distance	dist.	Highway	Hwy.
Distance	D	Hub & tack	H&T
Distance, horizontal	Dh	Inch	in.
District	Dist.	Inside diameter	ID
Ditch	dit.	Instrument	inst.
Drive	Dr.	Intersection	int
Driveway	drwy.	Iron pipe	IP
Drop inlet	DI	Irrigation pipe	irr.P

Junction . . . . jct.	Slope stake . . . . SS
Kilometer . . . . km	South . . . . S
Lane . . . . ln.	Spike . . . . spk.
Left . . . . lt.	Stake . . . . stk.
Manhole . . . . MH	Stand pipe . . . . SP
Marker . . . . mkr.	Station . . . . sta.
Maximum . . . . max.	Steel . . . . stl.
Measured . . . . meas.	Storm drain . . . . SDr.
Median . . . . med.	Street . . . . St.
Mile . . . . mi.	Structure . . . . str.
Millimeter . . . . mm	Subdivision . . . . subd.
Minimum . . . . min.	Subgrade . . . . SG
Minute . . . . min.	Tack . . . . tk.
Monument . . . . mon.	Telephone cable . . . . tel.C.
Nail . . . . N	Telephone pole . . . . tel.P.
North . . . . N	Temperature . . . . temp.
Number . . . . # or no.	Temporary bench mark . . . . TBM
Offset . . . . O/S	Top back of curb . . . . TBC
Original ground . . . . OG	Top of bank . . . . TB
Outside diameter . . . . OD	Top of curb . . . . TC
Overhead . . . . OH	Township . . . . T
Page . . . . p.	Tract . . . . tr.
Pages . . . . pp.	Transmission tower . . . . TT
Party chief . . . . PC	Turning point . . . . TP
Pavement . . . . pvmt.	Water line . . . . WL
Perforated metal pipe . . . . PMP	Water valve . . . . WV
Pipe . . . . P	Wing wall . . . . WW
Place . . . . pl.	
Plastic . . . . plas.	
Point . . . . pt.	

Common Abbreviations Used on Elevations

Point of intersection . . . . PI	Aluminum . . . . AL
Portland cement concrete . . . . PCC	Asbestos . . . . ASB
Power pole . . . . PP	Asphalt . . . . ASPH
Pressure . . . . press.	Basement . . . . BSMT
Private . . . . pvt.	Beveled . . . . BEV
Project control survey . . . . PCS	Brick . . . . BRK
Property line . . . . PL	Building . . . . BLDG
Punch mark . . . . PM	Cast iron CI
Railroad . . . . RR	Ceiling . . . . CLG
Railroad spike . . . . RRspk.	Cement . . . . CEM
Read head nail . . . . RH	Center . . . . CTR
Record . . . . rec.	Center line . . . . C or CL
Reference . . . . ref.	Clear . . . . CLR
Reference monument . . . . RM	Column . . . . COL
Reference point . . . . RP	Concrete . . . . CONC
Reinforced concrete pipe . . . . RCP	Concrete block . . . . CONC B
Retaining wall . . . . ret.W	Copper . . . . COP
Right . . . . rt.	Corner . . . . COR
Right of way . . . . R/W	Detail . . . . DET
River . . . . Riv.	Diameter . . . . DIA
Road . . . . rd.	Dimension . . . . DIM.
Roadway . . . . rdwy.	Ditto . . . . DO.
Rock . . . . rk.	Divided . . . . DIV
Route . . . . Rte.	Door . . . . DR
Section . . . . S	Double-hung window . . . . DHW
Sewer line (sanitary) . . . . SS	Down . . . . DN or D
Shoulder . . . . shldr.	Downspout . . . . DS
Sidewalk . . . . SW	

Drawing	DWG
Drip cap	DC
Each	EA
East	E
Elevation	EL
Entrance	ENT
Excavate	EXC
Exterior	EXT
Finish	FIN.
Flashing	FL
Floor	FL
Foot or feet	' or FT
Foundation	FND
Full size	FS
Galvanized	GALV
Galvanized iron	GI
Gauge	GA
Glass	GL
Glass block	GL BL
Grade	GR
Grade line	GL
Height	HGT, H, or HT
High point	H PT
Horizontal	HOR
Hose bibb	HB
Inch or inches	" or IN.
Insulating (insulated)	INS
Length	LGTH, LG, or L
Length overall	LOA
Level	LEV
Light	LT
Line	L
Lining	LN
Long	LG
Louver	LV
Low point	LP
Masonry opening	MO
Metal	MET. or M
Molding	MLDG
Mullion	MULL
North	N
Number	NO. or #
Opening	OPNG
Outlet	OUT
Outside diameter	OD
Overhead	OVHD
Panel	PNL
Perpendicular	PERP
Plate glass	PL GL
Plate height	PL HT
Radius	R
Revision	REV
Riser	R
Roof	RF
Roof drain	RD
Roofing	RFG
Rough	RGH
Saddle	SDL or S

Scale	SC
Schedule	SCH
Section	SECT
Sheathing	SHTHG
Sheet	SH
Shiplap	SHLP
Siding	SDG
South	S
Specifications	SPEC
Square	SQ
Square inch	SQ. IN.
Stainless steel	SST
Steel	STL
Stone	STN
Terra-cotta	TC
Thick or thickness	THK or T
Typical	TYP
Vertical	VERT
Waterproofing	WP
West	W
Width	W or WTH
Window	WDW
Wire glass	W GL
Wood	WD
Wrought iron	WI

Common Abbreviations Used on Plan Views

Access panel	AP
Acoustic	ACST
Acoustical tile	AT
Aggregate	AGGR
Air conditioning	AIR COND
Aluminum	AL
Anchor bolt	AB
Angle	AN
Apartment	APT
Approximate	APPROX
Architectural	ARCH
Area	A
Area drain	AD
Asbestos	ASB
Asbestos board	AB
Asphalt	ASPH
Asphalt tile	AT
Basement	BSMT
Bathroom	B
Bathtub	BT
Beam	BM
Bearing plate	BRG PL
Bedroom	BR
Blocking	BLKG
Blueprint	BP
Boiler	BLR
Bookshelves	BK SH
Brass	BRS
Brick	BRK

Bronze	BRZ	Exterior	EXT
Broom closet	BC	Finish	FIN.
Building	BLDG	Finished floor	FIN. FL
Building line	BL	Firebrick	FBRK
Cabinet	CAB.	Fireplace	FP
Caulking	CLKG	Fireproof	FPRF
Casing	CSG	Fixture	FIX.
Cast iron	CI	Flashing	FL
Cast stone	CS	Floor	FL
Catch basin	CB	Floor drain	FD
Cellar	CEL	Flooring	FLG
Cement	CEM	Fluorescent	FLUOR
Cement asbestos board	CEM AB	Flush	FL
Cement floor	CEM FL	Footing	FTG
Cement mortar	CEM MORT	Foundation	FND
Center	CTR	Frame	FR
Center to center	C to C	Full size	FS
Center line	C or CL	Furring	FUR
Center matched	CM	Galvanized iron	GI
Ceramic	CER	Garage	GAR
Channel	CHAN	Gas	G
Cinder block	CIN BL	Glass	GL
Circuit breaker	CIR BKR	Glass block	GL BL
Cleanout	CO	Grille	G
Cleanout door	COD	Gypsum	GYP
Clear glass	CL GL	Hardware	HDW
Closet	C, CL, or CLO	Hollow metal door	HMD
Cold air	CA	Hose bibb	HB
Cold water	CW	Hot air	HA
Collar beam	COL B	Hot water	HW
Concrete	CONC	Hot-water heater	HWH
Concrete block	CONC B	I beam	I
Concrete floor	CONC FL	Inside diameter	ID
Conduit	CND	Insulation	INS
Construction	CONST	Interior	INT
Contract	CONT	Iron	I
Copper	COP	Jamb	JB
Counter	CTR	Kitchen	K
Cubic feet	CU FT	Landing	LDG
Cutout	CO	Lath	LTH
Detail	DET	Laundry	LAU
Diagram	DIAG	Laundry tray	LT
Dimension	DIM.	Lavatory	LAV
Dining room	DR	Leader	L
Dishwasher	DW	Length	L, LG, or LNG
Ditto	DO.	Library	LIB
Double acting	DA	Light	LT
Double-strength glass	DSG	Limestone	LS
Down	DN	Linen closet	L CL
Downspout	DS	Lining	LN
Drain	D or DR	Linoleum	LINO
Drawing	DWG	Living room	LR
Dressed and matched	D & M	Louver	LV
Dryer	D	Main	MN
Electric panel	EP	Marble	MR
End to end	E to E	Masonry opening	MO
Excavate	EXC	Material	MATL
Expansion joint	EXP JT	Maximum	MAX

Medicine cabinet	MC
Minimum	MIN
Miscellaneous	MISC
Mixture	MIX
Modular	MOD
Mortar	MOR
Molding	MLDG
Nosing	NOS
Obscure glass	OBSC GL
On center	OC
Opening	OPNG
Outlet	OUT
Overall	OA
Overhead	OVHD
Pantry	PAN
Partition	PTN
Plaster	PL or PLAS
Plastered opening	PO
Plate	PL
Plate glass	PL GL
Platform	PLAT
Plumbing	PLBG
Porch	P
Precast	PRCST
Prefabricated	PREFAB
Pull switch	PS
Quarry tile floor	QTF
Radiator	RAD
Random	RDM
Range	R
Recessed	REC
Refrigerator	REF
Register	REG
Reinforce or reinforcing	REINF
Revision	REV
Riser	R
Roof	RF
Roof drain	RD
Room	RM or R
Rough	RGH
Rough opening	RGH OPNG
Rubber tile	R TILE
Scale	SC
Schedule	SCH
Screen	SCR
Scuttle	S
Section	SECT.
Select	SEL
Service	SERV
Sewer	SEW.
Sheathing	SHTHG
Sheet	SH
Shelf and rod	SH & RD
Shelving	SHELV
Shower	SH
Sill cock	SC
Single-strength glass	SSG
Sink	SK or S
Soil pipe	SP
Specifications	SPEC
Square feet	SQ FT
Stained	STN
Stairs	ST
Stairway	STWY
Standard	STD
Steel	ST or STL
Steel sash	SS
Storage	STG
Switch	SW or S
Telephone	TEL
Terra cotta	TC
Terrazzo	TER
Thermostat	THERMO
Threshold	TH
Toilet	T
Tongue-and-groove	T & G
Tread	TR or T
Typical	TYP
Unexcavated	UNEXC
Unfinished	UNF
Utility room	URM
Vent	V
Vent stock	VS
Vinyl tile	V TILE
Warm air	WA
Washing machine	WM
Water	W
Water closet	WC
Water heater	WH
Waterproof	WP
Weatherstripping	WS
Weep hole	WH
White pine	WP
Wide flange	WF
Wood	WD
Wood frame	WF
Yellow pine	YP

Bronze. . . . . BRZ
Broom closet. . . . . BC
Building. . . . . BLDG
Building line . . . . BL
Cabinet . . . . CAB.
Caulking . . . . CLKG
Casing. . . . . CSG
Cast iron. CI
Cast stone . . . . CS
Catch basin . . . . CB
Cellar. . . . . CEL
Cement . . . . CEM
Cement asbestos board. . . . . CEM AB
Cement floor. . . . . CEM FL
Cement mortar . . . . CEM MORT
Center. . . . . CTR
Center to center . . . . C to C
Center line . . . . C or CL
Center matched . . . . CM
Ceramic . . . . CER
Channel . . . . CHAN
Cinder block. . . . . CIN BL
Circuit breaker. . . . . CIR BKR
Cleanout. . . . . CO
Cleanout door . . . . COD
Clear glass . . . . CL GL
Closet. . . . . C, CL, or CLO
Cold air. . . . . CA
Cold water . . . . CW
Collar beam . . . . COL B
Concrete. . . . . CONC
Concrete block. . . . . CONC B
Concrete floor . . . . CONC FL
Conduit. . . . . CND
Construction . . . . CONST
Contract . . . . CONT
Copper. . . . . COP
Counter. . . . . CTR
Cubic feet . . . . CU FT
Cutout. . . . . CO
Detail . . . . DET
Diagram. . . . . DIAG
Dimension . . . . DIM.
Dining room. . . . . DR
Dishwasher. . . . . DW
Ditto. . . . . DO.
Double acting . . . . DA
Double-strength glass . . . . DSG
Down . . . . DN
Downspout. . . . . DS
Drain . . . . D or DR
Drawing. . . . . DWG
Dressed and matched. . . . . D & M
Dryer . . . . D
Electric panel . . . . EP
End to end. . . . . E to E
Excavate . . . . EXC
Expansion joint . . . . EXP JT

Exterior . . . . EXT
Finish . . . . FIN.
Finished floor . . . . FIN. FL
Firebrick . . . . FBRK
Fireplace . . . . FP
Fireproof . . . . FPRF
Fixture. . . . . FIX.
Flashing . . . . FL
Floor. . . . . FL
Floor drain . . . . FD
Flooring . . . . FLG
Fluorescent . . . . FLUOR
Flush. . . . . FL
Footing . . . . FTG
Foundation . . . . FND
Frame. . . . . FR
Full size . . . . FS
Furring . . . . FUR
Galvanized iron . . . . GI
Garage . . . . GAR
Gas. . . . . G
Glass . . . . GL
Glass block . . . . GL BL
Grille . . . . G
Gypsum . . . . GYP
Hardware. . . . . HDW
Hollow metal door . . . . HMD
Hose bibb . . . . HB
Hot air . . . . HA
Hot water . . . . HW
Hot-water heater . . . . HWH
I beam I
Inside diameter. . . . . ID
Insulation . . . . INS
Interior. . . . . INT
Iron I
Jamb . . . . JB
Kitchen . . . . K
Landing . . . . LDG
Lath. . . . . LTH
Laundry. . . . . LAU
Laundry tray. . . . . LT
Lavatory. . . . . LAV
Leader L
Length. . . . . L, LG, or LNG
Library . . . . LIB
Light. . . . . LT
Limestone . . . . LS
Linen closet . . . . L CL
Lining . . . . LN
Linoleum . . . . LINO
Living room . . . . LR
Louver LV
Main. . . . . MN
Marble . . . . MR
Masonry opening. . . . . MO
Material. . . . . MATL
Maximum . . . . MAX

Medicine cabinet	MC
Minimum	MIN
Miscellaneous	MISC
Mixture	MIX
Modular	MOD
Mortar	MOR
Molding	MLDG
Nosing	NOS
Obscure glass	OBSC GL
On center	OC
Opening	OPNG
Outlet	OUT
Overall	OA
Overhead	OVHD
Pantry	PAN
Partition	PTN
Plaster	PL or PLAS
Plastered opening	PO
Plate	PL
Plate glass	PL GL
Platform	PLAT
Plumbing	PLBG
Porch	P
Precast	PRCST
Prefabricated	PREFAB
Pull switch	PS
Quarry tile floor	QTF
Radiator	RAD
Random	RDM
Range	R
Recessed	REC
Refrigerator	REF
Register	REG
Reinforce or reinforcing	REINF
Revision	REV
Riser	R
Roof	RF
Roof drain	RD
Room	RM or R
Rough	RGH
Rough opening	RGH OPNG
Rubber tile	R TILE
Scale	SC
Schedule	SCH
Screen	SCR
Scuttle	S
Section	SECT.
Select	SEL
Service	SERV

Sewer	SEW.
Sheathing	SHTHG
Sheet	SH
Shelf and rod	SH & RD
Shelving	SHELV
Shower	SH
Sill cock	SC
Single-strength glass	SSG
Sink	SK or S
Soil pipe	SP
Specifications	SPEC
Square feet	SQ FT
Stained	STN
Stairs	ST
Stairway	STWY
Standard	STD
Steel	ST or STL
Steel sash	SS
Storage	STG
Switch	SW or S
Telephone	TEL
Terra cotta	TC
Terrazzo	TER
Thermostat	THERMO
Threshold	TH
Toilet	T
Tongue-and-groove	T & G
Tread	TR or T
Typical	TYP
Unexcavated	UNEXC
Unfinished	UNF
Utility room	URM
Vent	V
Vent stock	VS
Vinyl tile	V TILE
Warm air	WA
Washing machine	WM
Water	W
Water closet	WC
Water heater	WH
Waterproof	WP
Weatherstripping	WS
Weep hole	WH
White pine	WP
Wide flange	WF
Wood	WD
Wood frame	WF
Yellow pine	YP

Architectural Terms Commonly Found on Plans

Window Terms

Apron – A plain or molded piece of finish below the stool of a window that is put on to cover the rough edge of the plastering.

Drip cap – A projection of masonry or wood on the outside top of a window to protect the window from rain.

Head jamb – Horizontal top post used in the framing of a window or doorway.

Light – A pane of glass.

Lintel – Horizontal structural member supporting a wall over a window or other opening.

Meeting rail – The horizontal center rail of a sash in a double-hung window.

Mullion – A large, vertical division of a window opening.

Muntin – A strip of wood or metal that separates and supports the panes of glass in a window sash.

Sash – The part of a window in which panes of glass are set; it is generally movable, as in double-hung windows. The two side pieces are called stiles, and the upper and lower pieces are called rails.

Side jambs – Vertical sideposts used in the framing of a window or doorway.

Sill – Horizontal member at the bottom of a window or doorway.

Stool – A flat, narrow shelf forming the top member of the interior trim at the bottom of a window.

Stop bead – The strip on a window frame against which the sash slides.

Pitched Roof Terms

Flashing – Sheet metal, copper, lead, or tin that is used to cover open joints to make them waterproof.

Gable – The end of a ridged roof as distinguished from the front or rear side.

Louver – An opening for ventilation that is covered by sloping slats to exclude rain.

Ridge – The top edge of a roof where the two slopes meet.

Ridge board – A board that is placed on the edge at the ridge of a roof to support the upper ends of rafters.

Saddle – A tent-shaped portion of a roof between a chimney and the main part of the roof; built to support flashing and to direct water away from the chimney.

Valley – The intersection of the bottom two inclined sides of a roof.

Cornice Terms

Cornice – The part of a roof that projects beyond a wall.

Cornice return – The short portion of a cornice that is carried around the corner of a structure.

Crown molding – The molding at the top of the cornice and just under the roof.

Fascia – The outside flat member of a cornice.

Frieze – A trim member used just below the cornice.

Soffit – The underside of a cornice.

Stair Terms

Headroom – The distance between flights of steps or between the steps and the ceiling above.

Landing – The horizontal platform in a stairway.

Nosing – The overhanging edge of a stair tread.

Rise – The vertical distance from the top of a tread to the top of the next highest tread.

Riser – The vertical portion of a step.

Run – The horizontal distance that is covered by a flight of steps. Also, the net width of a step.

Stringer – The supporting timber at the sides of a staircase.

Tread – The horizontal part of a step on which the foot is placed.

Structural Terms

Anchor bolt – A bolt with the threaded portion projecting from a structure; generally used to hold the frame of a building secure against wind load. Anchor bolts may also be referred to as hold-down bolts, foundation bolts, and sill bolts.

Batt – A type of insulation designed to be installed between framing members.

Battens – Narrow strips of wood or metal used to cover vertical joints between boards and panels.

Beam – One of the principal horizontal members of a building.

Bridging – The process of bracing floor joists by fixing lateral members between them.

Camber – The concave or convex curvature of a surface.

Expansion joint – The separation between adjoining parts to allow for small relative movements, such as those caused by temperature changes.

Footing – The foundation for a column or the enlargement at the bottom of a wall to distribute the weight of the superstructure over a greater area to prevent settling.

Furring – Strips of wood or metal applied to a wall or other surface to make it level, form an airspace, or provide a fastening surface for a finish covering.

Girder – The main supporting beam (either timber or steel) that is used for supporting a superstructure.

Header – A wood beam that is set at a right angle to a joist to provide a seat or support.

Joist – A heavy piece of horizontal timber to which the boards of a floor or the latch of a ceiling is nailed. Joists are laid edgewise to form a floor support; they rest on the wall or on girders.

Monolithic concrete – A continuous mass of concrete that is cast as a single unit.

Plate – A structural member with a depth that is substantially smaller than its length or width.

Rake – Trim members that run parallel to the roof slope and form the finish between the roof and the wall at the gable end.

Reinforced concrete – Concrete containing metal rods, wires, or other slender members. It is designed in such a manner that the concrete and metal act together to resist forces.

Sill – A horizontal member that is supported by a foundation wall or piers, and which in turn bears the upright members of a frame.

Slab – A poured concrete floor.

Steam wall – That portion of a foundation that rests on the footing.

Studs – The vertical, slender wood or metal members that are used to support the elements in walls and partitions.

Vapor barrier – A material that is used to retard the flow of vapor or moisture into the walls or floors and thus prevent condensation within them.

Veneer – The covering layer of material for a wall or facing materials applied to the external surface of steel, reinforced concrete, or frame walls.

Additional Resources

This module is intended to be a thorough resource for task training. The following reference works are suggested for further study. These are optional materials for continued education rather than for task training.

Architectural Graphic Standards, The American Institute of Architects. New York, NY: John Wiley & Sons, Inc.

Print Reading for Construction, 1997. Walter C. Brown. Tinley Park, IL: The Goodheart-Willcox Company, Inc.

Reading Architectural Plans for Residential and Commercial Construction, 1998. Ernest R. Weidhaas. Upper Saddle River, NJ: Prentice Hall.

Figure Credits

John Hoerlein	104F04
Ivey Mechanical Company, LLC	104F09, 104F10
Holder Environmental Services	104F37

The NCCER makes every effort to keep these textbooks up-to-date and free of technical errors. We appreciate your help in this process. If you have an idea for improving this textbook, or if you find an error, a typographical mistake, or an inaccuracy in NCCER's *Contren®* textbooks, please write us, using this form or a photocopy. Be sure to include the exact module number, page number, a detailed description, and the correction, if applicable. Your input will be brought to the attention of the Technical Review Committee. Thank you for your assistance.

Instructors – If you found that additional materials were necessary in order to teach this module effectively, please let us know so that we may include them in the Equipment/Materials list in the Annotated Instructor's Guide.

Write: Product Development
National Center for Construction Education and Research
P.O. Box 141104, Gainesville, FL 32614-1104

Fax: 352-334-0932

E-mail: curriculum@nccer.org

Craft _____ Module Name _____

Copyright Date _____ Module Number _____ Page Number(s) _____

Description _____

_____

_____

_____

(Optional) Correction _____

_____

_____

(Optional) Your Name and Address _____

_____

_____

Index

Index

9 780131 091733